Bittersweet

THE STORY OF SUGAR

Peter Macinnis

ALLEN&UNWIN

First published in Australia in 2002

Allen & Unwin
83 Alexander Street
Crows Nest NSW 2065
Australia
Phone: (61 2) 8425 0100
Fax: (61 2) 9906 2218
Email: info@allenandunwin.com
Web: www.allenandunwin.com

National Library of Australia
Cataloguing-in-Publication entry:

Macinnis, P. (Peter).
  Bittersweet: the story of sugar.

  Bibliography.
  Includes index.
  ISBN 1 86508 657 6.

  1. Sugar—History. 2. Sugar trade—History. I. Title

633.6

Cover by Liz Seymour
Cover illustration: Theodor de Bry, *Collections peregrinatorium*, 1595
Chapter illustration: Jacques Dalechamps, *Historie générale des plantes*, 1615
Maps by Ian Faulkner
Typeset by Midland Typesetters
Printed by McPherson's Printing Group

10 9 8 7 6 5 4 3 2 1

CONTENTS

ABOUT
THE
AUTHOR

Peter Macinnis was born in Queensland, Australia, but grew up on Sydney's northern beaches, where he still lives—and where he is a somewhat obsessive bushwalker. After several years tending ledgers in the 1960s, he decided it was better to be a biologist than an accountant and obtained a bachelor's degree in botany and zoology. As well as being a husband and father of three, Peter has worked as many things: science teacher; education officer for the state department of education where he eventually became Principal Education Officer and number cruncher (as well as gaining a Masters in Education); and he was later an educator with both the Powerhouse and the Australian Museum before returning to the classroom for a number of years. In 1999 Peter began writing full-time for WebsterWorld, an online encyclopaedia, and continues his work there today.

Over the years, Peter has written or co-authored some twenty books, mainly for children or schools, as well as presenting talks on ABC Radio National since 1985, appearing on 'Science Bookshop', the 'Science Show' and, most frequently, on 'Ockham's Razor'. Peter has a large science web site (http://members.ozemail.com.au/~macinnis/scifun/index.htm) which has won many awards in the past few years. He has been twice awarded 'highly commended' in the Michael Daley Awards for Science Journalism, and in 2000 a children's book he co-authored received a Whitley award from the Royal Zoological Society of New South Wales.

Peter's work sits at the intersection of three interests: writing, science and education. His next book is about the history and development of rockets. He speaks several languages incredibly badly and has a black belt in bog-snorkelling.

ACKNOWLEDGMENTS

Many people and institutions helped me in the course of preparing this book. Most of them I found on the Internet, one way or another.

Among the people who helped, I have to count John Gilmore, who discussed James Grainger with me; Michael Sveda, who gave some insights into how his father discovered cyclamates; Alice Holtin in Arkansas, who filled me in on where to learn about southern US sugar cane; Bill Allsopp, also in Arkansas, an amazing educational thinker and fact-fossicker; Lan Wang, who gave me advance files of her scanning of the *Travels of Marco Polo*; and a whole host of people on the Science Matters list: Chris Forbes-Ewan, Margaret Ruwoldt, David Allen, Elizabeth May, Chris Lawson, Gerald Cairns, Tamara, Geoff ZeroSum, Sue Wright, Richard Gillespie and Stephen Berry among them.

Among the helpful institutions, I include the library of Manly Municipality, and its partners in the Shorelink library system, the main and branch libraries of the University of Sydney and the University of New South Wales, and also the excellent State Library of New South Wales.

I specifically acknowledge the Australian Government, which taxed all my photocopy charges, the books I bought while researching this, my notebooks and writing paper, my travel, the power that drove my computer, my software, and the shoes I wore out, and then, without having lifted a finger, had the temerity to level a tax on the finished book equal to the amount I get in royalties, and after that will filch half of my royalties as income tax. This service made it very much easier for me to understand the complaints of the sugar growers who had rapacious and parasitic tax-happy regimes to contend with.

At Allen & Unwin, Emma Jurisich, Jo Paul and Ian Bowring patiently prodded me in directions I did not wish to follow—at first—and made this a much better book by their persistence. Then Emma Cotter came on the scene, and showed what happens when a truly erudite editor is let loose on an errant author. When Emma was side-tracked by a brand-new baby, Narelle Segecic took over. It was a seamless transition as Narelle sorted the minor technicalities that arise in any book. I realise now that this editorial flair is the norm at Allen & Unwin. The remaining blemishes result from the recalcitrant author occasionally clinging to one of his *bons mots* contrary to good taste, editorial strictures, public order and the need for good writing.

Thanks also to Chris who has always put up with me, Angus, Cate and especially Duncan, who did some of the library chasing for me, and to Oz Worboys, who introduced me to story of the Kanakas long before I knew of William Wawn, or that he had sailed with my great-grand-uncle (who castigated him for bad language) in the Queensland labour trade. Mike Wright, the demon barber of Balgowlah, swapped philosophies, kept my head cool and prepared me to be photographed. My dentist Thomas Chai undid the damage sugar, and other dentists, had wrought upon my tooth and gave me a place to think.

Finding a name for a book is usually the hardest part. So I asked some friends, sent the best ones as I saw them to Ian Bowring, and then we settled on our best choice. We like it, but lying behind a simple title is a lot of careful thought, from people like Caroline Eising, Bruce Young and Doug Rickard, all in Brisbane, Amanda Credaro in western Sydney, Elizabeth May and Ian Jamie, both from Sydney University, my son Duncan Macinnis and my favourite son-in-law, Julian Ng, Janice Money in Darwin, Cathy Berchtold in Cincinnati, Ohio, Dan Heffron in Fort Myers, Florida, Jeri Bates in Victoria, British Columbia, John Bailey in Somerset, Kathleen Kisting Alam in Lahore and Nancy Baiter in Washington. Thanks, people!

My last-minute checks were ably assisted by the eagle eyes of my son, Angus, and of my good friend Jean Lowerison of San Diego.

I have only myself to blame, but all those people to thank.

NOTE ON
MEASUREMENTS
AND MONEY

As far as possible, I have converted measurements and monetary values to a common scale. Most of the masses given are approximations, converted to a notional 'ton', which is a mix of the metric tonne, the short ton and the avoirdupois ton, which may be divided into 20 hundredweights (cwt). With many of the gallon figures, it is unclear as to whether they should be taken as the modern US gallon (the old English wine gallon of 231 cubic inches) or the modern British gallon (the old English corn gallon of 268.8 cubic inches), or even the beer gallon of 282 cubic inches. 'A pint's a pound/the whole world round' say the Americans, even as the British and their Commonwealth once chanted 'a pint of pure water/weighs a pound and a quarter'.

Faced with the prospect of boring myself, my editor and my readers witless with voluminous conversions based on often unreliable assumptions about what was meant in the first place, I opted for a retreat into vagueness. I hope this vagueness will be appreciated.

For modern financial measures, I have used an indication of the value in American dollars, since most people seem to be able to translate this into their own currency. Many values, though, were in sterling, and I note here that one pound, three shillings and sixpence is generally written £1 3s. 6d. or as 23s. 6d. There are 20 shillings in a pound, and 12 pence in a shilling. Out of the goodness of my heart, I converted the two instances I encountered of guineas.

PREFACE

A treatise on sugar-refining (the dreariest subject I can think of) could have been given a more lively appearance.

Joseph Conrad, *Within the Tides*

Luckily, the story of sugar is about more than sugar refining. It is a sprawling tale that covers 9000 years, about 150 countries that produce beet or cane sugar, and a product that has influenced all our lives. In telling this story of many threads, the choices I have made from among the many issues help to show how sugar has been involved in world history, although I do not suggest that sugar caused the world to be as it is today. Rather, the human greed, frailty and misjudgment that have in part shaped our world have also operated around the profits to be made, one way and another, from sugar. If sugar had not been available, some other valuable item would have been squabbled over in a similar fashion.

Religion, in a number of guises, has played a part in the story of sugar as well. Sugar probably travelled from Indonesian Hindus to Indian Hindus, and later was made by Nestorian

Christians in Persia. After that, sugar followed Islam around the Mediterranean, so it could be discovered by the Crusaders who spread a taste for sugar across Europe. When Catholic nations and Protestant nations competed and warred in the New World and in the east, one of the things they fought over most was sugar.

Later, the wealth to be gained from growing and manufacturing sugar attracted men who sought power, some of whom justified their acts of inhumanity towards other humans, their slaves, in terms of a perverted version of their religion. It is, however, their selective interpretations that we must blame, not religion itself.

The story of sugar has very few heroes and many villains. The villains' consciences remained clear, because they believed that what they did was for the 'common good', as well as for the good of their pockets. Likewise, those who peddle strange 'scientific' claims about sugar and artificial sweeteners today believe that by doing so, they are saving lives. They may be wrong, they may be dishonest, but they seem to believe in their hearts that they are doing good.

Perhaps, before we judge any of sugar's historical figures too harshly, we need to look at ourselves today, and ask how we will be judged by our descendants. It is better to look at the evil men have done, in an effort to ensure that we do not repeat it, than to look upon past evils with a sanctimonious superiority. We are different, but it is doubtful we are that much better, for few things change as little as human nature.

INTRODUCTION

I t was cold in much of tropical New Guinea, 9000 years ago, as the last Ice Age was dying. Even close to the equator, the mountain ranges held glaciers and snow frosted the peaks in white all year round. Still, it was warmer on the coastal plain, and there was plenty of food growing in the jungle. There were worse places to live back then, even if some of the warmth came from volcanic activity.

Volcanoes have been good for humans, right across the sweep of islands that almost links Australia to Asia. The Australian continental plate is too thick and tough to crack or wrinkle as it slides north at the speed of a growing fingernail, so New Guinea and Indonesia take all the force. The islands are Australia's crumple zone, and all over the area volcanic outpourings and earthquakes are normal. The high mountains of New Guinea, pushed up by the progress of the Australian plate, wring water from the passing tropical clouds to feed the many rivers, and the plant growth in the rich soil is luxurious. To the west of New Guinea, most of Indonesia's islands are volcanic as well.

Volcanic rocks make fertile soil. In tropical areas, the

monsoon rains of the wet season leach out the minerals that plants need, but volcanoes replace the minerals just as quickly, covering worn-out soil with ash, or with lava that breaks down after a century or so into rich soil. It is an ideal place to farm, or to discover how to farm.

Volcanoes also provide raw materials that humans can use for tools. There is obsidian for blades to slice food, pumice for shaping wood, and thin basaltic flows and dykes that cool quickly to make sheets of a special stone that is almost like metal. In time, these sheets break up and the bits become cobbles in the rivers, ringing and rattling their way slowly down to the sea.

Picking these special stones out from the other cobbles is easy. The best blade-making stones are darker, with a glossier sheen, and when struck they ring like a bell. They also have holes in the surface, because the lava cooled fast, forming a glassy mass rather than crystals. As the pressure eased, gas bubbles expanded in the congealing rock, but they had no time to escape from the viscous lava before it cooled. Trapped in the rock, the bubbles pock the smooth surface of the rounded fragments and mark these fist-sized, rounded cobbles as special.

If they are struck hard enough on another boulder, at just the right angle, the stone will split to produce slivers of serrated blade, as sharp as razors. The cores that are left form choppers that make short work of trees and vines. Sometimes, a gas bubble made a hole right through the piece, and a leather thong could be threaded though the blade to hang around the neck. Polished and attached to timber handles, they made axes and adzes that were prized and traded over wide areas, being exchanged for pigs, shells, feathers, obsidian from far away, or even for brides. The volcanic stone blades made life easier, and gave neighbouring tribes more cohesion as they traded with each other.

Although the worst of the Ice Age was over 9000 years ago, it was still cold on the mountains, and so people lived close to the plain, hunting animals and gathering plants where they could be found. There was a pattern though, for the life of a hunter-gatherer is planned around the weather patterns that determine the food cycle.

In the tropics, where weather is driven by the monsoons, there is a wet season and a dry season, and in between there are different stages of drying out. Even in apparently steady conditions, migratory birds fly through on their way to or from Australia and so there are plenty of clues to judge the seasons by and, even near the equator, a wealth of micro-seasons. Even so, the wet season can't escape notice.

The term 'wet season' is a limp description of what really happens over several months of the southern summer. Not long after noon each day, the clouds roll in from the sea and pile up, the only warning of what is to come. When the rain starts, it is no gentle pattering shower but a sudden drumming, mind-numbing onslaught that terrifies and confuses. There is no thunder in the sky, just thunder on the trees as the rain sweeps in, thunder on the ground as it pours off the trees, and thunder in the heads of those trapped in the open by the rain, which descends at the rate of an inch (25 millimetres) or more in a mere 20 minutes. When you are out in rain like this, you only have one thought, and that is to get out of the downpour. There is only room for that one thought, so shelter is essential, at least until the rain passes, as quickly as it arrived.

THE BIRTH OF CANE FARMING

There are, and were, many staple foods on the island of New Guinea, most of them rather bland to European tastes; but growing wild in the jungle was a giant member of the grass family, as thick as bamboo, but not hollow like a bamboo stem. The stem was filled with juicy pith, and it could be chewed, sucked gently, then crushed between strong back teeth to release the delicious sweet sap inside. This was sugar cane, destined to be one of the first crops domesticated by humans, perhaps even the first agricultural crop, somewhere low on a New Guinea mountainside.

The easy way for a hunter-gatherer to harvest wild cane is to chop some of the long stems off at the base with a stone axe, and carry a bundle of long canes home over the shoulder, where they can be cut into smaller pieces with a smaller hand-held chopper, and shared out to chew on under shelter as the wet season rains beat down.

All that was needed for shelter was a simple open hut of the sort still seen all over the tropical Pacific. It has corner posts of timber, a roof of leaves or grass thatch, a palm-leaf mat resting on a bamboo platform where you can sit comfortably with your feet out of the mud, and places to store food, tools and other items that would otherwise be buried in the mud when the daily torrent starts. Perhaps, one day in the wet season, nine millennia ago, a piece of cut cane fell from a platform and was trodden into the mud by somebody hurrying to outrun the rain as it roared up the hill, stripping leaves from the trees as it came.

Like all the grasses, sugar cane has a jointed stem, and its leaves and branches come from shoots at each joint. In lawn grasses the joints may be hard to see, because the leaves form a sheath around the stem, hiding the inner workings, but sugar farmers around the world know that you grow new sugar cane by cutting off lengths with two or three joints, and placing these in the ground. They know because somebody told them, just as somebody else told *them*, in a line that stretches all the way back to that first discovery, somewhere in New Guinea.

Stone Age people may use a different tool kit, but they come with the same brain kit as agricultural and industrial peoples, and somebody, seeing unexpected sugar canes growing near a shelter, might have been tempted to pull them out of the ground. Seeing the cut section of cane that the shoots were springing from, an ingenious mind might then have made the world's first experiment by planting some cane lengths deliberately.

The fine detail of how it came about matters little. What counts is that, around 400 generations in the past, New Guineans were the first to discover a crop that was destined to change much of the world. Without the combination of blades of volcanic stone, rich soil and ferocious rain, the discovery might have taken longer—but it is enough that somebody found that small pieces of cane poked into the ground would sprout and grow more sugar cane—and it would be easy enough to learn this in the wet season, in a land of rich and sticky volcanic soil.

New Guinea is identified as the place where sugar cane was first cultivated because one of the original species found in later hybrid canes is still growing there. The other components of the hybrid cane appear to have come from India, and botanists assume that the New Guinea cane was carried and traded all the way to India, where the first hybrid canes came into being. The rest of the argument is complex botany, but suffice it to say that for half a century, botanists have regarded sugar cultivation as a New Guinean invention.

So why did the cane travel so far in early times? Even when they lack a common language, humans develop ways of communicating, of synthesising what linguists call creole languages and lingua francas, capable of transmitting complex ideas—and traders will happily transmit the idea that this sort of stick is nice to chew on, and so worth trading. Even in New Guinea, where people in neighbouring valleys often speak entirely different tongues, marrying-out occurs—this is a polite way of saying that women are 'traded', married off into other clans and tribes—and so methods and ideas travel from village to village.

A few men travelled more widely as traders of feathers, stone blades or other essentials of life. As they went, they would also help to spread new ideas about the sweet stick that grew when bits were poked into the ground. And that, of course, might have been the trigger to make other New Guineans start poking sticks in the ground, to see if they grew in the same way. Soon everywhere that sugar cane was found, people would have known the trick of putting bits in the ground. More importantly, sugar was being found in new places, as some of the extra bits were traded further afield, along with the key knowledge. And down in the lowlands, longer trading trips along the coast were possible using a creole language that has since blossomed as Bahasa Indonesia, the national language of Indonesia, from Malaysia to the western half of New Guinea.

The word 'creole' appears many times in the story of sugar. A creole language has a mixed but brief set of words which must often carry multiple meanings, and a recognisable syntax. These tongues arise whenever different racial groups come together. The Pidgin English of New Guinea uses words from many languages, but clearly has an Austronesian syntax, like Bahasa Indonesia. Creole languages also developed in Hawaii and many other sugar-growing regions. The wealthy planters of the Caribbean were called Creoles; the sugar cane that came to the Caribbean from the Mediterranean, the variety widely used across the world until the late 1700s, was called 'Creole'; and so were some of the mixed-race groups which arose in sugar-growing areas.

Somewhere, sometime later, perhaps in India, perhaps somewhere else, somebody found that if you boiled cane juice in a metal pan, added some ash or other alkali, scooped off the skin on the surface and boiled the juice some more, sweet crystals formed. The art of making sugar had been discovered, and a new industry was invented.

THE TWENTIETH-CENTURY SLAVE LINE

Just a teenager, I stood on a slight rise, watching the labour lines trudging across undulating ground. They were planting teak roots in a cleared patch of jungle on the coastal plain of Papua, setting the trees out in close rows. The procedure was simple and labour-intensive: lines of five men walked between stakes placed eight feet apart at each end of the ground they were filling that day. The stakes marked the rows where the trees were to go.

Each leader measured eight feet from the previous point, using his precisely cut pole, made a scratch and moved on to measure the next pole-length. Walking behind, the second man swung a pick to make a small hole and the next man, carrying a sack of teak roots cut early that morning, dropped a root beside the hole for the fourth man to poke into the ground. At the tail of the line, the last man jumped and landed, one bare foot to each side of the root, to firm the loose soil around the root.

It was a hot day at the start of the wet season, the perfect time to plant trees. The sun was almost overhead, and the clouds for the daily downpour were starting to mass up. The temperature was hovering around the century mark on the Fahrenheit scale, and the humidity was close to 100 per cent, but the labourers kept up a steady pace, back and forth, filling the cleared ground with future trees. As they went they chatted and laughed, but they never slowed their pace, except when they returned to the road. Then they stopped briefly to drink water or to cut a small piece of sugar cane from the lengths on the back of the truck, before the root-carriers took a new load of roots, and they set off again.

These were convict labourers, planting an export crop for a nation that did not yet exist, a crop that would one day provide the emerging nation of Papua New Guinea with foreign exchange. In Pidgin English, the creole language of the area, these were *kalabus slaves*. They had all committed violent crimes in the highlands and had been sent to serve their sentences on the coast in a gaol, which had somehow acquired the name 'calaboose', though with a local spelling—and a far cry from the original Spanish *calabozo*, which is a dungeon. Creole languages have few rules, and words mean whatever you like, so these were kalabus slaves.

The forestry officer standing with me explained that just one old man supervised the kalabus slaves, but they accepted that they were in the kalabus for a reason, they knew they would have good food and shelter that night, and the work was less boring than sitting in Boumana gaol. They knew they were a long way from home, and they had little idea of how to get there, so they were content to work out their sentences.

Soon the daily rains would come, just after midday, and the men would all scramble onto the back of the truck and return to their nominal prison. When the rain stopped they would tend the sugar cane, bananas and other plants in the small prison garden. While they called themselves kalabus slaves, it was an example of how an adopted word had mutated when it was taken into Pidgin, he said.

'But they really *are* slaves, aren't they?' I asked. 'I mean, they're made to work, and they get no pay . . .'

'Not really,' said the man. 'They get a bit of money, more than they'd get in gaol, but that's beside the point. By the time these trees are thinned, this'll be an independent nation, so when the thinnings are made into veneer, they and their kids will reap the profits, not us. Besides, you'll see convicts planting trees in Australia as well—it's the normal thing.'

Then he gave me a piece of advice that older men have been giving younger men for as long as humans have used forced labour. 'Watch how you go,' he said. 'You're very new here, and you're full of noble thoughts, but this is what we do and how we do it, so don't go saying too much, because some people won't like it. Now the rain's coming, so let's go.'

I turned my back on one of the last slave lines in the world, and walked back to the truck, with the sweat pouring off me. As I walked, I chewed on a piece of thick sweet sugar cane, a traditional New Guinea garden delicacy that one of the convicts had cut for me with the heavy, razor-sharp machetes they used. At the back of my mouth, the sweet juices commenced an attack that 40 years later would demolish a left molar tooth and leave me in a dentist's chair, musing about Shakespeare.

IN THE DENTIST'S CHAIR

Twenty years later, half the teak trees were thinned to make second-rate veneer, giving the other trees more room to grow. Another 20 years, and the mature logs were coming out of the plantation. At the same time, after half a lifetime's neglect, sugar cane and bad care had finally done for my molar tooth, so in early 2001 it was coming out as well. Abscesses, root canal therapy, bad dentistry and capping had left just a remnant that must be removed, slowly and in very small pieces, so an implant could be inserted in its place.

I am, let me admit it, a total coward around needles and dentists. Many years ago, I found that lying back and doing a complex calculation, like the cube root of seventeen, took my mind off sharp things being introduced to my mouth. There was a problem on this day, however, because after an hour or so, having got my answer to three decimal places, I found I was losing track of the numbers, and while the dentist had lost count of the tooth pieces he was by no means finished. So I cast around for something else to occupy my mind.

I am also, let me admit it, a total slob around research, falling back on the methodology of New Electronic Brutalism whenever possible, using electronic assistants to find what I want. A few weeks earlier, I had been looking into the ways in which we use the word 'pie'. I knew Shakespeare called a magpie a Maggot Pie, and while I was relieved to find that Maggot was an old form of Margaret (so Mag Pie was just the sister to Jack Daw), my curiosity had been aroused about pies in general.

I had turned to one of my brutal tools, a monster text file of all of Shakespeare's plays, to search out how the Bard used 'pie' at different times. That led me to *The Winter's Tale*, and the plans the Clown lays to make a warden pie, for which he lists his needs:

> Let me see: what am I to buy for our sheep-shearing feast? Three pound of sugar, five pound of currants, rice—what will this sister of mine do with rice? But my father hath made her mistress of the feast, and she lays it on.

As I lay back in the chair, having my mouth beaten into submission, trying to plan a light essay on the pies of various sorts, the Clown's sugar came back to me. Sugar was the main source of my present dental predicament, but there was something odd about the Clown's list. As I understood it, sugar came to England from the West Indies, and Britain colonised the islands after Shakespeare was dead. So how could there have been any sugar around in Shakespeare's time? Didn't they use honey?

That set me wondering, and that was how this book came to be, because once I was out of the chair I went data-digging, and found that Shakespeare uses the word 'sugar' seventeen times in the plays and sonnets to mean sweetness, so his audiences must have understood the term. Still, sugar did not dominate, and 'honey' appears 52 times in his works in a similar role.

In time I learned that by 1600, sugar from the Mediterranean, from Africa and from islands in the Atlantic was being traded all over Europe. Sugar had travelled a long way from a clearing in New Guinea, through Indonesia, into India, Persia, Egypt and Palestine. On its travels, people had learned how to work with it, even though they had no idea of where it originated, but sugar was by no means yet the maker and breaker of fortunes and empires that it would become.

By Shakespeare's time, people had learned that making sweet tastes is a marvellous way to gather the money that gives power. Ever since, the story of sweetness has been the story of money and power, and the special kinds of corruption that follow from money and power in large amounts. Here follows the story of sugar, what it made, and what was made of it.

ANTI-GONORRHOEAL MIXTURE

Take of copaibe ½ oz., spirits of nitric ether ½ oz., powdered acacia 1 drm., powered white sugar 1 drm., compound spts. of lavender 2 drms., tinc. of opium 1 drm., distilled water 4 oz.; mix. Dose, a tablespoonful three times a-day. Shake before using.

Daniel Young, *Young's Demonstrative Translation
of Scientific Secrets*, Toronto, 1861

1
THE
BEGINNINGS

Sugar cane is a member of the grass family. In botanical language it is *Saccharum officinarum*, a name given to it by Linnaeus himself—Carl von Linné, the inventor of our modern classification system. This, and five related *Saccharum* species, are placed in the Andropogoneae tribe, along with sorghum and maize. Like other grasses, sugar cane has jointed stems and sheathing leaf bases, with the leaves, shoots and roots all coming from the stem joints.

The world's scriptures have few references to sugar. Sugar rates no mention in the Quran (which, as we will see later, is significant), and while both Isaiah 43:24 and Jeremiah 6:20 refer to 'sweet cane', which some people think *might* mean sugar cane, there are a number of other candidates. If we assume that sugar was intended where the Bible's translators wrote of 'sweet cane', then the line in Jeremiah, 'To what purpose cometh there to me incense from Sheba, and the sweet cane from a far country?', tells us that sugar cane did not grow around Palestine in Old Testament times. The problem here, as ever, is that we are in the hands of translators who interpreted the Old Testament in terms of their own understandings and assumptions about the past.

The only world religious leader who makes any specific reference to sugar is Gautama Buddha. His words were written down some time after his death, so there may have been some interpolations, but he was probably familiar with at least some form of sugar cane. Buddha was, after all, born about 568 BC, at a time when the sugar cane was probably known and grown in India.

The set of instructions known collectively as the Buddhist rule of life, the *Pratimoksha*, defines *pakittiya* or self-indulgence as seeking delicacies such as ghee, butter, oil, honey, fish, flesh, milk curds or *gur* (a form of sugar) when one is not sick. As this particular rule was laid down by Buddha himself, it suggests he was at least aware of sugar. As well, when Buddha was asked to allow women to enter an order of nuns, he likened women in religion to the disease *manjitthika* (literally, 'madder-colour', after the red dye called madder) which destroys ripe cane fields, and which is caused by *Colletotrichum falcatum*. This fungal disease of cane still exists, going by the common name red rot.

There are other Indian references to sugar from this period, but the exact sort of sugar meant is never clear. Still, it seems there were cane crops large enough to suffer disease in Buddha's time, around 550 BC, and a Persian military expedition in 510 BC certainly saw sugar cane growing in India. The army of Alexander the Great reached India around 325 BC, and Nearchus, one of Alexander's generals, wrote later of how 'a reed in India brings forth honey without the help of bees, from which an intoxicating drink is made though the plant bears no fruit'. We now take this to mean that he saw sugar cane and sugar juice, but not sugar itself. This comment has often been used to argue that sugar cane was taken to Egypt by Alexander at about this time, but there is no evidence for that.

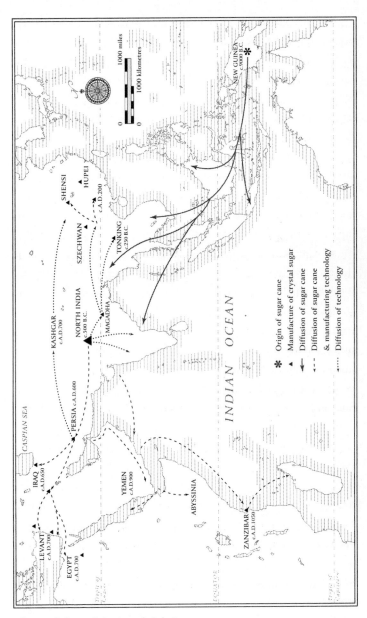

The spread of sugar before it reached the Europeans.

Around 320 BC, a government official in India recorded five distinct kinds of sugar, including three significant names: *guda*, *khanda* (which is the origin of today's 'candy') and *sarkara*. If the date is correct, this would appear to be evidence that sugar was being converted into solids in some way before 300 BC, so perhaps Nearchus did see sugar after all. Remember the *guda* and *sarkara*, because we will meet them again.

By about 200 BC sugar cane was well known in China, although it is possible that it was only chewed as cane. There is a record from AD 286 of the Kingdom of Funan (probably Cambodia) sending sugar cane as a tribute to China. Some 500 years earlier, in the late part of the Chou dynasty, it was recorded that sugar cane was widespread in Indochina. It is also possible that sugar cane was being grown at Beijing around 100 BC, though it is hard to tell how successful this would have been, 40 degrees north of the equator. We do know that sugar cane was already on the move, and could have reached Africa at about this time, and perhaps Oman and Arabia. The important move of sugar (as opposed to sugar cane) onto the world stage seems not to have come until around AD 600, when the cultivation of sugar cane and the art of sugar making was definitely known in Persia, at least to the Nestorian Christians who lived there.

Sugar cane was an important crop in India long before this, and nobody seems to know quite why it took so long to reach Persia (modern Iran). Perhaps it has to do with the irrigation that the cane needs in dry areas. In AD 262, Shapur I, a Sassanid king, made a dam at Tuster on the Karun (Little Tigris) River in Persia to irrigate surrounding areas by gravity feed relying on the height of the waters behind the dam. The water was eventually used to irrigate cane, and the ruins of the irrigation works are still there.

The centralised system of authority that was the Persian

empire would have allowed the development of large-scale irrigation schemes. Major irrigation schemes anywhere, like the terraced rice fields of Java and Bali, the fields along the Nile and Australia's irrigation areas, all rely on a central authority to provide organisation and an imposed peace.

A community of Nestorian Christians was certainly making good sugar in Persia around AD 600. If the art of sugar making had now been perfected, this could explain why sugar suddenly took off. The crop and its product had only spread slowly up until then, so clearly something happened: either there was a change in the method of growing cane or in the methods of extracting sugar, or maybe there was a change in the nature of the cane. Then again, as many writers have suggested in the past, perhaps sugar just followed the spread of Islam, once Islamic forces had defeated the Sassanid dynasty of Persia.

THE TRUE INVENTORS OF SUGAR?

All over the world, the word for sugar seems to come from the Sanskrit *shakkara*, which means 'granular material'. We find words like the Arabic *sakkar*, the Turkish *sheker*, the Italian *zucchero*, the Spanish *azúcar*, the French *sucre* and, of course, the English 'sugar'. It is *sukker* in Danish and Norwegian, *sykur* in Icelandic, *socker* in Swedish, *suiker* in Dutch and *zucker* in German. Yoruba speakers in Nigeria call it *suga*, Swahili East Africa calls it *sukari*, Russians call it *sachar*, Romanians say *zahar* and the Welsh call it *siwgwr*—and when you allow for the Welsh pronunciation of 'w' (rather like 'oo' in 'book'), the pattern is retained.

Bahasa Indonesia is one of the few languages where this pattern does not apply. Here, the name for sugar is *gula*, although when biochemists in Indonesia speak of 'sugars' as a group the name they give them is *sakar*. The Arabic origins of that are clear enough, but that expert among experts on Malay etymology, R. O. Winstedt, said in his early twentieth-century dictionary of the Malay language that he could see the Sanskrit origins of *gula* just as easily. But he said so without knowing that sugar cane originated on the island of New Guinea, at the far end of the Indonesian archipelago. Tradition then had it that sugar cane originated in India or China, and Winstedt was an old man when its true origins were worked out.

Few Europeans know much of the immense Hindu influence on Java and Bali. The various Javanese empires traded with India over many centuries, and perhaps sugar in a prepared form was first traded to India from Java, not the other way around. In that case, when the art of sugar making was learned in India, a Sanskrit word similar to the established Indonesian word would have been applied to the product the Javanese knew as *gula*. So *gula* would have given its name to the Indian *gur*, rather than the other way around.

Why would the Indians call it *gur*? The European linguists say the sugar came out of the boiling-pan as a sticky, treacly ball, and *gur* is a Sanskrit word for a ball. All the other lands heard about sugar as *shakkara*. Why would Indonesia alone have a different name for sugar, unless it was their word to begin with?

Perhaps the Indians who brought the Hindu religion to Java came from a place where *gur* or *gula* or even *guda* was used in preference to *shakkara*, but it would be unfair to rule out an Indonesian origin for the first refining of sugar to crystals. Bronze drums were well known in the archipelago, and ironware *could* have been traded there quite early on, so an Indonesian origin of sugar is at least possible.

I admit this is speculation, and the question has to remain an open one. The earliest records of sugar crystals seem to come from India, where a Sanskrit manuscript dating from about AD 375 refers to *sito sarkara churna*, but this 'powdered white sugar' may have been formed simply by drying *gur*. Manuscripts from that period are hard to date, but certainly by the fifth century AD, and quite possibly much earlier, we seem to see the first descriptions in Sanskrit of the preparation of sugar as we know it from cane. On the other hand, old manuscripts only rarely survive, so who can say what Indonesian records are missing?

THE SACKING OF DASTAGERD

The year AD 622 was a key year for three religions: the growing Islam under its prophet Muhammad, the Zoroastrianism of the Persians under the Emperor Chosroes II, and the Christianity of Byzantine Rome, based in Constantinople where Heraclius was emperor. Chosroes held the upper hand, and when an unknown Meccan sent him a letter, calling upon him to acknowledge Muhammad as the apostle of God, Chosroes rejected the invitation and tore the epistle to pieces. He had the Roman Empire on the run; what need had he of such an upstart as Muhammad, who called himself Prophet?

A bit of background: Phokas, Emperor of Byzantium AD 602–610, had been deposed by Heraclius, and Byzantium was tottering. A group called the Avars was attacking the European part of the Roman Empire. Chosroes was quietly taking Asia Minor, bit by bit, using the pretext that he was avenging Maurice, who had been deposed and murdered by Phokas.

This vengeance claim rang a little hollow once Heraclius took the throne from Phokas, but Chosroes already held Syria when Heraclius was crowned, and he continued to advance in the name of Zoroastrianism—with Jewish, Nestorian and Jacobite

Christian allies. In short, the Middle East in the seventh century was as troubled as it is in the twenty-first century.

In 615, when Chosroes held most of the Middle East, Muhammad predicted in the 30th surah of the Quran, called Ar-Rum ('The Greeks'), that the forces of Byzantium (known to Muhammad as the Greeks) would be victorious over Persia. At the time this was a daring and improbable claim, for just a year earlier Chosroes had written scornfully to Heraclius from Jerusalem: 'From Chosroes, the greatest of all gods, the master of the whole world: To Heraclius, his most wretched and most stupid servant: you say that you have trust in your Lord. Why didn't then your Lord save Jerusalem from me?' The future for Byzantium was looking bleak.

When Muhammad moved to Medina in 622, Heraclius was just setting out on a series of campaigns that paradoxically would open the way for Islam to advance. The Roman emperor led his troops on 48-mile marches in 24 hours, out-fought and out-thought the Persians and tore apart their empire. He forced Chosroes into defeat after defeat and retreat after retreat, until in 627 Chosroes was deposed, to 'die in a dungeon' five days later, after seeing his eighteen sons killed. Peace was made and Rome regained all of its lost territory. Heraclius was freed at last of the burdens of war, but the two mutually weakened empires were ready to be taken over by forces which flocked to the once-obscure Meccan upstart. Byzantium and Persia had worn each other out and, in the last eight years of his rule, Heraclius saw all of his regained provinces fall to the Arabs.

When Chosroes fled his royal palace at Dastagerd, the Roman forces found extremely fine pickings. Among the loot they found aloe wood, pepper, silk, ginger and sugar, described as 'an Indian delicacy'—which by then it almost certainly was not. All the same, this is an important clue, because it suggests that while sugar making might have been unknown beyond the Persian empire, sugar itself had been heard of, and seen.

Islam gained substantially from the Byzantium–Persia conflict, because the successful Quranic predictions like the one in 'The Greeks' were hailed as evidence that Muhammad was a true prophet. This gave Islam a dominant position in the Arab world, and it was now free to move into the power vacuum. Islam was on the move, and right in its path lay the places in Persia where sugar was being prepared by Nestorian Christians. Soon after, the Muslims would acquire many other areas where sugar cane would grow.

THE WANDERING SUGAR CANE

This foray into Middle Eastern history has taken the story slightly ahead of itself, however. From its early beginnings as a crop for chewing and sucking on, the sugar cane had spread first with coastal traders. The people of South-East Asia and the islands, like their Polynesian descendants, were excellent navigators, launching themselves out into the Pacific on trading journeys 4000 years ago. These people all speak languages of the Austronesian group, and they probably originated somewhere around Taiwan before spreading as far as Easter Island, New Zealand, Fiji, Hawaii, Tahiti, Indonesia and the Philippines, and all the way across the Indian Ocean to Madagascar.

We know where the early seafarers went in the western Pacific, because we find the signs of their travels in remnants of obsidian, a volcanic glass carried from New Britain to New Ireland some 15 000 or 20 000 years ago. They left adzes, used for making dugout canoes, at sites which appear to be 13 000 years old. Three thousand years ago, New Britain obsidian was travelling as far as Sabah in north Borneo. Then there are the marsupials that were taken from New Guinea, and perhaps Halmahera, to stock small islands, presumably for hunting purposes. The archaeological record shows the sudden appearance of both cuscuses (cat-sized nocturnal possums) and wallabies on the island of Gebe 10 000 years ago. But at least 30 000 years earlier, humans made sea crossings beyond any sight of land, as the first people travelled to Australia. Even when the Ice Age had lowered sea levels 100 metres or more, the crossings were still daunting.

English speakers and Europeans in general rarely know that Asia and the Pacific had brave and skilled seafarers long before the time of Leif Eiriksson. They have no idea that Chinese fleets visited East Africa before Vasco da Gama entered the Indian Ocean, or that giant bronze drums made by the lost wax method at Dongson, close to Hanoi, were carried to places like Bali well before the Christian era, as traders slipped from island to island, crept around coasts, and occasionally used seasonal winds to make longer ocean crossings. And at some stage, somewhere along the way, these seafarers carried sugar cane to wherever it would grow. Only later did it spread to Persia, the Mediterranean and across the Atlantic.

Taxation records clearly show taxes imposed between AD 636 and AD 644 on sugar grown in Mesopotamia. The sugar captured at Dastagerd in 627 had probably originated in Persia, but it was still an Indian product as far as most people were concerned. That was about to change, however. In 632 the prophet Muhammad died, and soon afterward the forces of Islam commenced their amazing expansion. In 637 the Persians were defeated at Kadysia, which meant the final collapse of the Sassanid empire, and now the Muslim forces were in a position to discover sugar (as opposed to sugar cane) for themselves.

Certainly we can reject the legend that Marco Polo brought the art of sugar refining back from the east in the thirteenth century, because he commented on the similarities between the Chinese and Egyptian methods of making sugar. All we can say for certain is that sugar cane came from New Guinea, was traded along a variety of coasts, spread inland, hybridised with other species in India and travelled some more, and that somewhere between Indonesia and Persia, about 1500 years ago, somebody discovered how to make sugar from the juice of the sugar cane. From that point, sugar and the technology it demanded began to travel, and as the technology spread it started changing things. Garden cane for chewing could spread by simple diffusion, but the idea of sugar technology was different, because people were able to carry that across continents and oceans. It was an idea as much as a crop.

Wherever it was encountered, sugar was highly valued. It is hardly surprising that Columbus took sugar cane to the West Indies or that the First Fleet carried sugar cane from the Cape of Good Hope to Australia. Later, when refined sugar reached New Guinea, it was named *siuga* in Pidgin English. Sugar had come all the way back home, but along the way it had changed remarkably, from a sticky sweet sap sucked from the cane to pure white crystals in paper bags.

More to the point, by the time sugar came home again it had helped change the world. It had proved a troublesome crop—it had made fortunes, caused rebellions, battles and bloodshed, made and broken empires, led to the enslavement and death of millions and, in the process, to the transplanting of blocks of humanity around the world, taking 20 million Africans to the Americas, Japanese and Chinese to Hawaii, Indians to the West Indies, the Pacific, Mauritius and Natal in South Africa, and Pacific Islanders to Australia, all in the interests of making other people rich.

FLATHONYS

Take mylke, and yolks of egges and ale, and draw hem
thorgh a straynour, with white sugur or black; and melt
faire butter, and put thereto salt, and make faire coffyns,
and put hem into a Nowne till þei be a little hard; þen
take a pile, and a dish fastened thereon, and fill þe coffyns
therewith of the seid stuffs and late hem bake while. And
þen take hem oute and serue hem forthe, and caste Sugur
ynough on hem.

Harleian ms 279: fifteenth-century cookbook

2

THE
SPREAD
OF SUGAR

The sugar trade began slowly because it was competing with an ancient and established honey trade. There is a Neolithic painting from the Araña cave at Bicorp in Spain which is usually interpreted as a man robbing a bees' nest on a cliff (though the waist and hips look more feminine than masculine). Until about 650 BC, hives were robbed, but after this, apiculture (bee-keeping) became more common.

The honey makers had no monopoly on sweetness. Sorghum had been domesticated in Ethiopia around 3000 BC, before spreading to the rest of Africa and also to India somewhere between 1500 and 1000 BC. Fig and date syrups were both common around the Mediterranean Sea but honey was used widely, from India to northern Europe. Honey, which is used to make mead, is *medd* in Welsh and *meodu* in Old English, while it is *mádhu* in Sanskrit and *med* in the Czech tongue, the similarities suggesting a single origin. The Romans called honey *mel*, showing the same consonant shift we see in *guda/gula*. They also boiled grape juice to increase the concentration of sugars to a point where nothing could live in the thick fluid.

Although syrup is obtained from maple trees and sugar beet, and many fruits contain sugars, honey was the main sweetener for most early societies. The Romans of the first century AD were very partial to sweet tastes. Virgil wrote Georgic verse on the art of bee-keeping as a source of sweetness. Pliny actually mentioned sugar, noting that the largest pieces were the size of a hazelnut, and that sugar was reserved for use as a medicine (*ad medicinum tantum usum*).

In AD 698 Ina, king of Sussex, allowed rents to be paid in honey, and his people drank mead, but by 950, when Hywel the Good in Wales set out a code which extended special protection to the brewers of mead, sugar cane was already growing in Spain and throughout the Mediterranean. Sugar was poised to become an essential commodity.

SUGAR AND ISLAM

While there is no mention in the Quran of sugar, the Prophet was extremely clear about fermented drinks of the sort that Nearchus had noted in India almost a millennium earlier. Alcohol was forbidden to those who held the faith; for refreshment, good Muslims were restricted to sweet drinks with no alcoholic content, which must have made sugar a highly desirable discovery. For 800 years after the fall of the Sassanids, the production of sugar generally followed Islam, not because of any great advances in technology, but because Islam covered a large area in which existing technologies could be combined and spread.

An intoxicating brew could be made from sugar, just by leaving a sugar solution or even the ordinary juice lying around to ferment. Preventing fermentation required cunning. The cane juice had to be boiled down to a syrup so sugary that no organism could ferment it—a cordial—which could later be watered down to make a refreshing drink. The traditional cordial *sekanjabin*, a sweet mint drink, was mentioned by the author al-Nadim as far back as the tenth century. A variety of other drinks were stored as syrups in medieval Islamic society. (This method remains in use to this day, with modern Coca-Cola and similar drinks being distributed as syrupy concentrates for post-mixing.)

The spread of Islam brought a *Pax Arabica*, permitting trade over a wide area, and also allowing scholars to travel, to observe, and to spread ideas, all the way from China to the Atlantic, from Norway and Russia to the Indonesian archipelago, and south into Africa. And with no disrespect intended, there were almost certainly those Muslims who adhered to the letter of the words of the Prophet and avoided wine, while leaving their sugary drinks to stand for a day or two to add a bit of zing. Here is how Sir John Mandeville, who claimed to have travelled to Palestine, described the practice in about 1366:

> . . . Saracens that be devout drink never no wine. But some drink it privily; for if they drunk it openly, they should be reproved. But they drink good beverage and sweet and nourishing that is made of gal-lamelle and that is that men make sugar of, that is of right good savour, and it is good for the breast.

In reality, Mandeville did not travel in the area at all, and often told the most outrageous lies about Jews and Muslims. Nonetheless, the interesting intoxicating effect of fermented cane juice was known in many societies.

The mid-600s saw a massive expansion of Islam and sugar. Islamic forces took Cyprus in 644, and even in the first half of the century there were references to sugar cane cultivation in Syria, Palestine and Egypt. By the end of the century sugar had spread all around the Mediterranean, wherever irrigation was available. By 700 the water wheel or *noria* had been brought into use in the crushing of cane, and a period of consolidation followed. Islamic forces reached Spain by 711, but it was only in 755 that the ruler Abd-ar-Rahman I felt things were peaceful enough to send an expedition to the eastern Mediterranean for sugar cane and other plants for his garden in Cordoba.

On a domestic scale, existing technology was adequate for the new crop. The cane was cut into short lengths and crushed under an edge-runner mill, where a stone wheel rolled on its edge around a circular groove. This extracted some of the juice, and more was obtained from a beam press or a screw press. The edge-runner mill had long been used to crush olives, nuts and mineral ores, while the presses were commonly used on grapes and olives. At first, Mediterranean sugar production was essentially a family industry.

At the same time, Muslim travellers visiting tropical areas invariably found sugar cane growing. In 846 a traveller called Ibn Kordadhbeh saw sugar being made in Java, while in 851 an Arab master mariner called Soleiman noted the presence of sugar cane on Madagascar. In the 900s Muslim expansion began to slow. They lost control of Crete in 960 and Cyprus six years later, but their cane fields and sugar mills remained. We know Sicily had mills by this time, because Ibn Hauqal wrote in 950 that the 'Persian reed' was growing on the banks of the rivers and streams around Palermo, and that juice was obtained by feeding cane into pressing mills. By the eleventh century, sugar cane was being grown in Morocco and Tunisia.

In 1060 the Normans invaded Sicily and by 1090 the balance of power in the Mediterranean was changing fast: the Normans controlled Sicily, and the Arabs lost Malta in the same year. The first Crusade began in 1099, and soon enough,

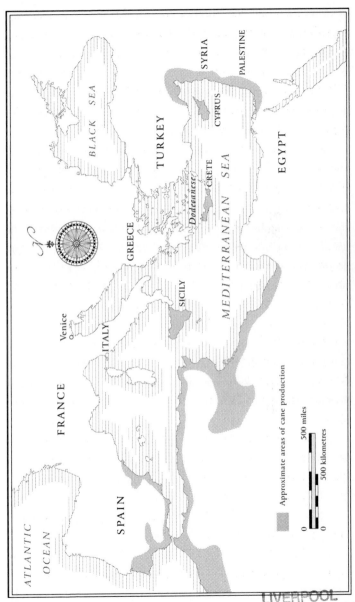

Sugar in the Mediterranean after AD 700.

returning Crusaders were spreading the word in Europe about sugar and sugar cane. Sugar was imported into northern Europe, initially as a medicine, and then as an addition to food and drink.

The medical uses of sugar have a venerable tradition. Buddha declared that it was no sin for a sick person to ask for *gur*, and Pliny had noted sugar's medicinal uses. The prophet Muhammad recommended dates, which have a very high sugar content, as medicine. In the thirteenth century, a theological debate in the Catholic church dealt with the status of sugar. Was it medicine, nourishment, or just for pleasure? Thomas Aquinas took the view that those who resorted to sugar during Lent did so for health reasons, not for nourishment—and in 1353 a French royal decree required apothecaries to swear never to use honey when sugar was prescribed.

However, things change. By 1581 Abraham Ortelius, a Flemish cartographer, would comment that 'what used to be kept by apothecaries for sick people only is now commonly devoured out of gluttony'. Even so, the French kept the expression 'like an apothecary without sugar' to indicate a state of utter desperation or helplessness. We still use sugar in medicines today, but in the Mary Poppins mode of improving the taste rather than as a cure. The sugar market has changed vastly from what it was around 1100, when the Crusaders first came across sugar as they travelled to a holy war of looting and pillaging.

THE FOUR CURSES OF SUGAR

Notwithstanding Mr Conrad's concerns about the dreariness of sugar-refining treatises, we need at least a basic understanding of the process that reduces sugar cane to a commodity which can be sold. We also need some insight into the fact that sugar is a special crop with special problems. These problems are the four curses of sugar.

From the sett, the short length of cane planted in the ground, to refined sugar is a costly process. The ground must be cleared, the setts have to be planted, and the field kept free of weeds until the cane is high enough not to be shaded out. After the first cane is cut, some fifteen months later, the root stock will produce several more ratoon crops at yearly intervals, but after several years the ratoons need to be replaced.

The cut cane is hauled to a crushing mill, where the juice is squeezed out. In early forms of refining, the sticky sap would be collected in containers before being boiled down, while a worker skimmed the scum off the top, until crystals formed: sugar crystals which were lifted off with a wooden spatula, leaving a rich brown molasses behind. This sugar was brown and sticky, not unlike the Indian *gur*, and far from the white, free-running crystals we know today.

This simple technology was known in Islamic society from the Umayyad period, which ended in AD 750. When the Crusaders arrived in Palestine they were happy to cash in, expanding production for export to Europe. The remnants of the sugar mills of Tawahin Al Sukkar may still be seen in Jericho, along with the remains of the aqueduct which brought the water from Ain Duyuk to power the mill.

Workers squashed the cane in a water-driven mill, and then pressed the flattened cane to extract the juice, which was boiled down in copper vessels, presumably with the addition of some ash. The product was placed in wicker baskets or earthenware containers to dry and harden. With the Crusaders in charge of the operation, people from all over Europe now had access to the technology, and to the cane.

Growing sugar as a commercial crop was rather different from growing a small garden or household supply. In its simplest form, the crushed-out juice is a mixture of water, sugar and a number of impurities, mainly suspended materials, fats, waxes and proteins, some of which need to be destroyed as soon as possible by boiling. The nuisance proteins are enzymes in the cane which convert the sucrose, or cane sugar, to 'invert sugar', a mix of glucose and fructose, once it is cut. Sucrose is a disaccharide, a molecule made of these two simple sugars or monosaccharides. Both glucose and fructose also taste sweet, but they lack some of the properties that sucrose offers when foods are being made.

In some parts of the world, high-fructose corn syrup may be used in foods, but bakers, confectionery and cereal manufacturers overwhelmingly prefer sucrose because of its bulking, texture and browning characteristics, so it is a disaster if the sucrose inverts in the cane before it is processed. As a rule, the juice needs to be boiled within 16 to 24 hours of cutting, in order to destroy the enzymes and so save the sucrose. Adding an alkali—either ash or lime—brings the impurities out of the solution as a precipitate called filter mud, while some of the other material forms a layer on the surface that needs to be skimmed off. Further boiling then leads to sugar crystals.

Once sugar growing became a larger-scale activity, a system of powered rollers was needed to crush the cane, as well as a channel to deliver the juice to the first of a series of metal boiling pans, and a suitable alkali for addition to the first pan of juice. Fuel was needed to heat the juice, land to grow the cane, and reserves to keep the growers going for fifteen months until the first crop matured. They needed transport to get the cane to where it would be crushed, iron hoops and timber staves for barrels, coopers and carpenters, a blacksmith to make or repair tools, animals and animal drivers, people to gather feed for the animals and to grow food for the workers, and much more. This is the first curse of sugar: it is capital intensive. To make a profit from a mill, once it was established, there is a certain minimum size of crop.

Where a number of people tried to share a central mill, however, tensions could arise when different farmers had cane ready for processing at the same time. Because the enzymes in the cane juice begin to do their work as soon as the cane is cut, immediate treatment is required. The need for speed in processing the crop is the second curse of sugar.

The mature sugar cane stem is about 75 per cent water and 15 per cent sucrose, while the rest is mainly fibre. To obtain a ton of sugar, more than seven tons of cane must be cut and hauled to the mill to be crushed, and five tons of water evaporated. This is the third curse of sugar, that it is hungry for fuel. All too often, sugar planters would destroy the forests around their plantations to obtain fuel.

Sugar making in the past was labour intensive. The fourth curse of sugar arose when slavery became a cheap solution to the labour problem.

A CROP BECOMES AN INDUSTRY

The Crusaders' Palestinian sugar operations produced only small amounts. (These days it would be called a boutique industry.) Sugar production for the European market did not become commercial until Crete came under Venetian control in 1204. Sugar was initially so valuable that the household records of England's King Henry III refer to the procurement of three pounds of sugar in 1226 as a special item, yet less than 400 years later Shakespeare's Clown regarded that same amount as a normal purchase. Still, the Clown was about to be robbed, so perhaps Shakespeare was signalling that a fair amount of money was involved.

In 1280 a cargo of sugar on 600 camels, bound from Egypt to Baghdad, was captured by Mongols under Hulagu, the grandson of Genghis Khan. The Middle Eastern sugar trade had become a major item after sugar production in Persia was largely wiped out following the death of Mostasim, the last Abassid Caliph, killed by Hulagu's Mongols in 1258. From the end of the thirteenth century, the sugar sold to Europe and the Middle East came from the Mediterranean area, which was just a single ship voyage away from the ports of northern Europe.

This was a situation the seafaring Venetians were gradually able to exploit. Just as the great Muslim traders had done before them, the Venetians travelled, explored and reported on what they saw. Some historians doubt that Marco Polo (1254–1324) really made it to China, but if he did not, then he was careful to gather intelligence from those who *had* been there, and so we should trust him when he writes of Fu-ju (Fukien):

> They have an enormous quantity of sugar. From this city the Great Khan gets all the sugar that is used at Court, enough to represent a considerable sum in value. You must know that in these parts before the Great Khan subjected it to his overlordship, the people did not know how to prepare and refine sugar as is done in Egypt. They did not let it congeal and solidify in moulds, but merely boiled and skimmed it, so that it hardened to a kind of paste, and was black in colour. But after the country had been conquered by the Great Khan, there came into the regions men of Egypt who had been at the Court of the Great Khan, and who taught them to refine it with the ashes of certain trees.

Around 1285 Polo reported that sugar was being made in China, India and east Africa, but he noted only the production of palm wine in Java. Around this time, Muslims captured the rest of Palestine, largely closing European access to its sugar. This stimulated production in other parts of the Mediterranean.

The first recorded commercial importation of sugar into England was in 1319, but during the fourteenth century sugar remained rare and expensive at two shillings a pound. As a measure of its rarity, Geoffrey Chaucer, a wealthy courtier and diplomat as well as a writer, makes only five references to sugar in his *Canterbury Tales*, written in about 1387. It would never be so costly again.

Production in Palestine, Egypt and Syria declined in the fourteenth century as Cyprus, Crete and the western Mediterranean took over as the main centres. The fall-off in Egypt may have been due to Mameluke misrule, as some writers claim. It may also have been due to population losses in the middle of the century from the Black Death, which disrupted both labour-intensive sugar growing and the maintenance of Nile irrigation systems, since Nile sugar needed 28 separate irrigations, all managed by hand. This drop in production gave Venice and Genoa the opening they needed.

THE NAVIGATORS, SUGAR AND SLAVERY

We can often see patterns in history that suggest inevitability. Before Robert Louis Stevenson wrote *Treasure Island*, children's books almost always had stern morals attached, but new worlds of fiction opened up after that book appeared, including the adult adventure fiction of Arthur Conan Doyle and others, catering for the *Treasure Island* generation as they came of age. It seems as though all the small bits gradually added together, like raindrops in a failing dam, until eventually there was a monstrous outpouring.

In retrospect, Renaissance navigation looks like a similar outburst. The factors that came together included better materials for sails and cordage, a better understanding of timbers, new ways of rigging boats so they could sail into the wind, better compasses—and a need to sail into new areas. That need came about in part because of Islamic sea power and piracy in the Mediterranean, and in part because there were good trading reasons for venturing into the Atlantic. It was known that there were islands out there, some inhabited by people who could be conquered, and some free of humans, ready for the taking. So whatever the cause, the fifteenth century saw ships sailing off in all directions.

The first Atlantic islands were discovered about 40 BC. The date of an expedition to the Canaries is fixed by a description from the learned Juba II, King of Numidia (roughly equivalent to modern Algeria). Later, this visit was also recorded by Pliny. In 1334 a French ship was driven to the Canaries in bad weather, and later returned. By about 1352, Catalan and Mallorcan missionaries are believed to have visited the Canaries. After that, the Canary Islands and their population came under Spanish rule.

When the Portuguese arrived on Madeira in 1421 they found no inhabitants, but excellent timber, and land for all. Portuguese settlers poured in and by 1432 the first sugar was being refined there. Serious exports really only began somewhere between 1450 and 1460, when Madeiran sugar was first carried to England and Flanders. By 1500, Madeiran sugar was available all over Europe. Prince Henry ('the Navigator') granted a licence for a water mill on Madeira in 1452, in exchange for one-third of the sugar produced there, but water-powered sugar mills remained rare.

It was at this point that slavery, the fourth curse of sugar, came into play. Slavery might have been stopped when the Papal Bull of 1454, *Romanus Pontilex*, allowed the conversion of native populations to Christianity. Unfortunately, that same bull gave Portugal a monopoly over the lucrative African slave trade. The Vatican said it was opposed to slavery in principle, but while Pope Pius II banned the enslavement of baptised Africans in 1462, and three other fifteenth-century popes condemned slavery in various terms, no practical measures were taken, and the slave trade began its exponential growth.

By 1470 sugar refineries had been established in Venice, Bologna and Antwerp to process imported raw sugar, setting the pattern that would hold through into the nineteenth century of cheap raw sugar being sent to metropolitan centres where expensive refined sugar would be made. The later restrictions on colonial refining were mainly to retain as much of the profit as possible in the 'home' countries, while the role of the colonies was to supply cheap sugar. Sadly, cheap sugar required cheap labour, and that meant slaves.

There was another side to the question of where refining should take place. With the rather limited forms of refining in use until the twentieth century, and with long, slow sea voyages, the sticky crystals would lock together into a solid mass under the humid conditions at sea. This made refinement near the final market a necessity in order to get the best quality sugar. Until the steamship became common, sugar had to be refined close to the point where it was consumed. This is why Australia had a sugar refinery to process imported raw sugar 20 years before it grew successful sugar cane crops.

Toward the end of the fifteenth century, sea traffic on the Atlantic was increasing. Christopher Columbus was sailing the ocean, taking sugar cane to Madeira, where his wife's mother grew it. Around 1480 the first sugar cane was planted in the Canary Islands and, further south, the Portuguese settled the island of São Tomé in 1486. Originally using the children of

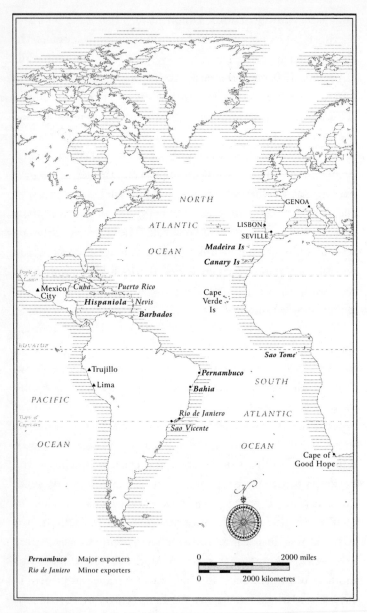

Sugar in the Atlantic from AD 1500 onwards.

Jews expelled from Spain as slaves on São Tomé, they soon afterwards brought in black slaves, and planted sugar cane there. As these and other new centres came into operation, using more efficient production methods, sugar became a bigger and bigger player in the world's politics.

By 1496 Madeira was shipping 1700 tons of sugar to Venice, Genoa, Flanders and England, and there were about 80 mills on the island. A few years later, an attack of 'worm' reduced sugar production, and by 1600 grape vines had largely taken over until, in 1852, an attack of mildew blighted the vines, and led people back to sugar. Less than 20 years later, Madeira's sugar production was back to a third of what it had been in 1496, but in another reversal it was found possible to re-establish the grape vines. Only a small sugar production survived, but by now sugar cane had reached the New World, where it would become a crop of world importance.

A number of factors boosted the slowly developing sugar industry in the Americas. One was the division of the world between Portugal and Spain under the Treaty of Tordesillas in 1493, which kept the Spanish out of Africa, at the same time allowing the Portuguese into the coastal areas of Brazil. Having once gained a toehold, the Portuguese were able to penetrate well beyond the line set down under the treaty, with the Spaniards safely on the other side of the Andes.

The result of the treaty was to give Portugal—and other European countries who ignored the Portuguese 'monopoly' and moved in on the slave trade that Portugal claimed—rights in Africa to obtain and transport slaves to the Spanish dominions. While trade with foreign ships was not officially allowed in the Spanish colonies, once the slave ships were actually at anchor in the Spanish colonial ports, commercial transactions took place, and so other countries became more aware of prospects in the West Indies. It also led to the establishment of a string of English, French and Portuguese forts in western Africa that became the seeds of the African colonial possessions that would leave permanent traces on the African map.

Events in the eastern Mediterranean also helped boost sugar in the New World. For instance, when the Turks took Syria in 1516, the Syrian sugar industry collapsed. Between 1520 and 1570 the Turks conquered Cyprus, Crete, Egypt, the Aegean and much of the north African coastal belt. Sugar prices rose as those industries crumbled, and that made it easier for adventurers to get the money they needed to grow sugar elsewhere. The Mediterranean sugar industry may also have been influenced by climatic change, with the 'Little Ice Age' of 1550–1700 taking perhaps 1–2°C off temperatures as well as reducing rainfall. Brazil, on the other hand, had plenty of rain, all the fuel it needed close at hand, and an excellent labour supply.

By the time Syria fell, Brazilian sugar was already being taxed when it arrived at Lisbon. King Manuel had ordered that a sugar master be sent to Brazil, and the first *trapiche*, an animal- or human-powered mill, had been built on San Domingo. The slaves on São Tomé had been growing sugar for around 30 years, but most of the New World slaves were used in the search for gold. That proportion would soon change as it became apparent that gold was easier to win by cultivating the sugar cane than by digging mines. By 1550 the pattern of slavery and sugar had been set.

FOR THE SMALL POX . . . AND BITING OF A MADD DOGG

Take of sage and Rhue of each a handfull, one spoonfull of Pewter scrapt very small, 3 heads of garlick halfe a pd: of running treacle, put all into a quart of strong Ale and put all into a pipkin with a liner and past it up very close. Let it stand in a gentle fire until half be consumed. If the smallpox fall suddenly flat and turne blew then give it to a man 5 or 6 spoonfulls at a time, to a child 2 or 3. If the party be in grave danger give this in 2 or 3 hours. The treacle here meant runy like syrup and is not above 8 pence a pound.

Bradford ms, no date

3
SUGAR IN THE
NEW WORLD

How should we remember Christopher Columbus? Others had sailed the Atlantic Ocean before him, settling islands and meeting the locals. St Brendan, the Irish monk, crossed the Atlantic in a leather cockleshell, and Leif Eiriksson reached North America as well. When the English cod fishers reached the cod-rich banks of Newfoundland in 1480, the secretive Basques of southern France and northern Spain were already there, twelve years before Columbus reached the West Indies.

THE ITALIAN NAVIGATOR

Columbus cannot claim to have worked out that the world is a globe, because the Greeks knew all about the spherical Earth 2000 years before he set sail. They used evidence including the Earth's shadow appearing as a circle on the moon in a lunar eclipse; the way a ship's hull disappears before the masts as the ship stands out to sea; and even the relative time of day or night an eclipse of the sun or the moon was seen in different parts of the Mediterranean.

Some time before 200 BC, Eratosthenes measured the globe's size using noonday sightings of the sun at Aswan and Alexandria to estimate the circumference of the Earth. We know that he was within about 4 per cent of the true size, though scholars made a botch of it for many years. But Eratosthenes had more to say, and he said it plainly: 'If the extent of the Atlantic Ocean were not an obstacle, we might easily pass by sea from Iberia to India, keeping in the same parallel.' People had always travelled east by land to the Indies and China. On foot it was a long way, and their estimates always set the Indies a greater distance to the east than they were. The other problem was that Eratosthenes had measured the world in a unit called a stadion, and nobody knew quite how big a stadion was, so they had to guess as well as they could. Sadly, their 'modern' stadion was too small by one part in four, and that made the Earth smaller than it is. That led them to think the Indies even closer to Europe, barely over the horizon from the Atlantic islands.

It was only logical that somebody would try sailing to the Indies, just across the Atlantic to the west. All Columbus did was to try what Eratosthenes had suggested. Of course Columbus made a small mistake, that scholars' botch I mentioned before. Knowing roughly how big the planet was thought to be, and that all the best spices and gold and sugar came from the Indies, Columbus worked out how far it would be if he went west instead of east.

Perhaps we should recall Columbus for the deed that would change the West Indies forever—taking sugar cane from La Gomera in the Canary Islands to Hispaniola on his second trip in 1493. By that simple act, he probably did more than anybody else to shape the Americas of today. Many of the events that moulded the New World had their roots in the cane fields, one way or another.

The conquistadors who came after Columbus thought the cane was a native of the West Indies, since it could be seen growing in Indian villages visited by Europeans for the first time. Then Spanish priests reported that sugar cane grew also in the Philippines, and made up complicated stories of traders carrying it across the Pacific. The simpler explanation was that when people saw and tasted the marvellous cane, it didn't take them long to grow their own. Nor did it take others long to trade bits of the cane to villages further afield, and in this way sugar cane swept through the West Indies faster than the conquistadors could expand. Sadly for the Indians, though, the conquistadors were spreading fast.

GOD'S FIRST PRIEST IN THE INDIES

Bartolomé de Las Casas was a Dominican missionary on Cuba in 1514, the first priest ordained in the Indies. He had a grant of land to help him do God's work, with a hundred local Caribs attached to the land as serfs. Within three years, this gentle and decent man was horrified to see how the Caribs sickened and died when they were forced to work, and how they were slaughtered by the Spanish colonists when they revolted. It would be better, he told the Court in Spain, to bring in Africans who were inured to such labour. He believed the choice to be the lesser of two evils, but he thought the lesser evil to be hardly an evil at all, for the African slaves in Spain seemed happy enough, and they were clearly better able to carry out hard work in the mines. In any case, the mines would soon be worked out, and the slaves could then be freed to a life of agriculture.

Instead, the trade grew ever larger, and African slaves soon began to outnumber their white masters. In 1530 there were 3000 slaves on San Domingo, and just 327 Spaniards. By 1547, de Las Casas was Bishop of Chiapa in Mexico, but he resigned his bishopric to return to Spain to campaign against the slave trade—the lesser evil he himself had suggested for the benefit of the Caribs and their peaceful neighbours, the Arawaks. It took three and a half centuries, five normal full spans of three-score years and ten, to abolish what he thought he could end as quickly as he had started. That same five life spans would be some 35 working lives of the African slaves, who never lived as long as free men, because slavery was cruel and brutal in the extreme.

At one stage, de Las Casas had great hopes that the King would listen to what he had to say, but early in 1555 that chance slipped away as Charles V began to fear for his own soul. He abdicated the throne and devoted the rest of his life to prayer in a little house next to the monastery of Yste, seeking salvation through prayer rather than by doing good deeds. If the king had struck against slavery then, perhaps half of all those who would eventually be hauled across the seas and worked to death might have been spared. Bartolomé de Las Casas gave the rest of his days to the fight, but he never again got close to winning.

His countrymen were not necessarily cruel to slaves *per se*. Like the English and most other nationalities at the time, they were just cruel to other humans in their power, making no particular discrimination between free men and slaves. It was normal to treat other human beings badly. People were still being burnt at the stake, and public executions, hangings, drawings and quarterings were a common enough ending for many a free man, and even a few women. Horrible things were done to the slaves, but mostly because they outnumbered the whites, and needed to be terrorised to remind them they had no rights. After all, no rational man would beat, murder or maim a slave, any more than he would beat a cart horse beyond what was needed to make it work, for slaves and horses were property, and both cost money to replace.

And slaves were not all that profitable at the best of times—Adam Smith knew that when he wrote *The Wealth of Nations* and, according to him, so did the ancients:

> The experience of all ages and nations, I believe, demonstrates that the work done by slaves, though it appears to cost only their maintenance, is in the end the dearest of any. A person who can acquire no property, can have no other interest but to eat as much, and to labour as little as possible. Whatever work he does beyond what is sufficient to purchase his own maintenance can be squeezed out of him by violence only, and not by any interest of his own. In ancient Italy, how much the cultivation of corn degenerated, how unprofitable it became to the master when it fell under the management of slaves, is remarked by both Pliny and Columella.

Sadly, not all slave owners in the Americas were entirely rational. Drunkenness from the abundant cheap rum, irritability caused by the tropical heat, even madness, led to actions that took a great toll of slaves' lives.

But far more numerous than the deaths from cruelty were the deaths from epidemic disease, always a problem when populations come together from distant places. Each group tends to carry a range of diseases to which they have developed a degree of immunity, and yet they have absolutely no immunity to the diseases carried by other groups.

The Arawaks, Caribs and other locals had the worst of it, because they were susceptible to both African and European diseases. The Spaniards arrived with influenza, pleurisy, measles, tuberculosis and whooping cough, which immediately began a savage attack on the locals. Then in 1516, smallpox arrived to ravage the local populations even further—a major reason for introducing slaves from elsewhere. The Indios, said the Spaniards, were a feeble bunch who died all too easily; clearly, they were not up to the work.

Long before the germ theory of disease, it made a sort of sense to blame the work. The Indians were sick; it could not always have been like this, so the different factor must be that the Indians were now being made to work. The solution was just as obvious: bring in Africans who were from an agricultural culture and known to be able to work in the heat, and set them to work. The slaves and the ships that brought them across the Atlantic carried diseases such as malaria, yellow fever, dengue fever, hookworm and schistosomiasis. In no time at all these diseases attacked the Indians—but found equally good targets in the white populations, among both the plantation owners and the indentured workers brought by the English when they moved in.

And then there was syphilis. Nobody seems to know whether syphilis came to the New World from the Old with Columbus's sailors, or if they took it back to Europe, but there was a lot of it about in those times, and in the end it made men mad, and cruel. Most of the cruelty, though, was seen in its time as for the good of the slaves, because it made better servants of them, and achieving that was their owners' first duty.

Sugar was lurking in the background, the threat that would last long after the mines were found and worked out. It was sugar that shaped the slaves' world, just as it reached forward into our time, shaping the world we know. Without sugar, there might only have been a minor slave trade; with sugar, the slave trade drove European commerce and development.

THE FIRST SUGAR SLAVES

Sugar and slavery seemed to go hand in hand. Without slaves, there might not have been a sugar industry—but without sugar, there would not have been as many slaves. On the other hand, slavery was around long before sugar came on the scene. Slavery is often thought of today as a distinctly American and Caribbean phenomenon, but slaves were being brought into fourteenth-century Crete and Cyprus to work in the Mediterranean cane fields even before the Black Death cut the local labour force. These were Greek, Bulgarian and Turkish prisoners of war, and Tartars. And going on old place names that survive today, historians believe that Morocco probably used slaves in the sugar industry around 1400.

In ancient times slavery was commonly the fate of those defeated in war, as the Hebrews found when Nebuchadnezzar came to call. There may have been slaves in what is now Libya, about 6000 BC, and there were certainly slaves in ancient Egypt. When Hammurabi of Babylon made the first code of laws, between 1800 and 1750 BC, there was a special provision that any who helped a slave escape, or who harboured a fugitive, should be put to death.

That high point in our western civilisation, Athens at the time of Socrates, Plato and Aristotle, around 450 BC, could only function through having enough slaves to free up citizens for study and thought. In the Athens of Pericles, perhaps two-thirds of the city's population were slaves of one sort or another. In the first book of his *Politics*, Aristotle explains that: 'It is clear, then, that some men are by nature free, and others slaves, and that for these latter slavery is both expedient and right.' At another point, Aristotle explains to his readers that the ox is the poor man's slave, and in general treats slavery as a natural phenomenon.

The Romans also relied on slaves, brought to Rome in huge numbers by their victorious armies, or brought from further afield in smaller numbers. During Rome's greatest period, from 50 BC to AD 150, there were probably half a million slaves brought into Rome and its empire each year—and why not? The Romans admired Greek learning and followed its teachings, and later Europeans followed the Romans in a manner that can only be called slavish. When Romans criticised slavery, it was the cruel treatment of some slaves that worried them, not the institution of slavery itself. If the great Aristotle said slavery was acceptable, that was an end to the matter.

While the slaves of the last thousand years have mainly been African in origin, black slaves were a minority in Roman times, though one is shown in a mosaic at Pompeii. Black slaves are also depicted on Greek vases.

Islam recognises the prophets of the Old Testament. Muslims, like Jews and Christians, were aware that Ham had seen Noah both drunk and naked, and that Ham was accursed for this. In Christian and Muslim tradition, Ham's descendants turned black, and the Hamitic peoples were destined for slavery, or so the slavers claimed. Something that had been decreed to be the will of both God and Allah, and was also profitable, would almost inevitably become almost unstoppable.

The rise of Christianity made little difference to slavery, even though the church mildly encouraged manumission, the act of freeing slaves from servitude. Pope Leo the Great declared in 443 that no slave could be a priest, and about the only time slaves did well was under a law of 417, which allowed that if a Jew bought a male slave and made him undergo circumcision, the slave was, by that act, made free. It seems there was a greater prejudice against Jews than there was against slaves.

Slavery remained the norm in Europe in the Dark Ages, but towards the end of the first millennium, the rent-paying serf slowly replaced the slave in much of northern Europe. The Domesday Book of 1086 reveals some 25 000 slaves in England, but by 1200 they had disappeared, perhaps because improved agricultural tools and methods meant there was more profit to be had from a serf.

Without the development of a large market for sugar, slavery may well have died out, but around the Mediterranean in the fourteenth and fifteenth centuries slavery remained common. Muslims, Christians, Jews and pagans all preyed on each other and took slaves where they could. It must have seemed as though a benevolent deity had placed others there so that people might avoid the embarrassment of having to enslave their own kind. Africans, of course, were fair game to all.

Soon, rather than slaves being a by-product of war, raids were being mounted just to catch them. The Muslims, in particular, took slaves from among the pagan Slavs in such large numbers that this gave rise to the word 'slave'. These Slavic slaves were widely traded around the Mediterranean until the Muslim capture of Constantinople in 1453 meant this source was largely lost to Europe. Still, it hardly mattered, because at the far end of the Mediterranean the Portuguese were already beginning to venture into the Atlantic, and along the African coast, returning with black slaves.

Slaves from Guinea had been sold in Europe from about 1250, brought overland to the Mediterranean coast by Moorish traders. At the same time the Portuguese were reaching out westwards. In 1320 the Canary Islands, known to the ancients as the Hesperides or the Fortunate Isles, were rediscovered by the Genoese navigator Lanzarotto Malocello. The inhabitants of these rediscovered isles were less fortunate, as many of them were promptly enslaved.

In the early 1400s African slaves became more common all around the Mediterranean, although most were house servants rather than field hands or industrial workers. That was soon to change. European expansion continued, and after the fall of the Moroccan port of Ceuta in 1415, the Portuguese were able to gather better information about Africa through first-hand information from overland traders who had travelled far to the south.

The first slaving incident of that era was the Portuguese seizure of a Moroccan ship in 1425 with 53 black men and three black women aboard, all from Guinea. They were liberated from the torments of Moorish enslavement and sold into benevolent Christian slavery—at a great profit to the liberators. How overjoyed the Portuguese must have been, to both fill their pockets and do a Christian duty at the same time! Unfortunately, such joint religious and economic opportunities were rare.

The odd thing is not so much that slaves were brought from Africa in Portuguese ships, but that it took so long for Europeans to work their way around the African coast. The first circumnavigation of Africa happened some time before the death of Herodotus in 425 BC, because he tells us about it in Book 4 of his *Histories*, even as he discounts the tale:

> These men made a statement which I myself do not believe, though others may, to the effect that as they sailed on a westerly course round the southern end of Libya, they had the sun on their right—to the northward of them. This is how Libya was first discovered to be surrounded by sea . . .

The Libya of Herodotus was the whole of the African continent. Anywhere south of Aswan on the Nile River, at the height of summer, the sun will be seen to the north at noon. So if later readers followed Herodotus and rejected the circumnavigation tale, it could only be because they were rejecting the evidence.

All the same, Cape Bojador, just south of the Canaries and almost on the Tropic of Cancer, was regarded as the safe limit for Portuguese travel. Further south, it was asserted, white sailors would turn black (as the locals had clearly done), and monsters, liquid flame and serpents would prevail. Perhaps those who had already ventured there spread the tale to frighten off others.

Antão Gonçalves was one of two captains who sailed to Cabo Branco (Cap Blanc or Râs Nouâdhibou), five degrees south of Cape Bojador and well into the tropics, in 1441. In bringing twelve black Africans before Prince Henry of Portugal, he was following an ancient tradition that stretched back at least to the Romans, who paraded captives before their emperor. It was a tradition that would be honoured by Columbus when he took Arawak 'Indians' back to Spain, and by the English when they took Pocahontas to meet King James I and again when they took two Australian Aboriginals, Bennelong and Yammerawannie, to London to meet King George III, and later still, Jemmy Button to meet yet another king.

One of Gonçalves's captives spoke Arabic, and negotiated his release in exchange for assistance in acquiring further slaves. In 1442 Gonçalves brought back another ten captives, while in 1443 Nuno Tristão captured fourteen men from canoes. He later increased this to 29, and the stage was set for the new slave trade. On 8 August 1444 a revenue officer named Lançarote de Freitas and his men landed near Lagos with 235 African slaves, all captured during a visit to what is now Mauritania.

The Azores were certainly known by 1351, when they first appeared on a map. The Portuguese claimed the islands in about 1431 and colonised them in 1445, just in time to start taking advantage of the new trade in African slaves. Within a few years, about a thousand slaves had been carried to the Azores and Madeira to cultivate sugar cane. The pattern of sugar and slavery was thus set not in the Indies but in the Atlantic Ocean.

The importation of slaves into the New World began slowly. A decree of 1501 forbade any slaves born in Spain being taken to the Indies, and the ban applied also to Jews, Moors and 'New Christians' (Jews who had been converted to Christianity). All the same, by 1502 the first black slave had arrived and in 1504 five white slaves (that is, Muslims) were brought in. They came not to work in the fields but to dig for gold in Spanish mines.

The Governor of Hispaniola, Nicolás de Ovando, opposed bringing in slaves from overseas. His grounds for this had nothing to do with a concern for fellow humans; rather, that these ingrates repaid their masters' efforts in importing them by taking every chance to run away. Worse, they encouraged the local slaves to do the same! They were a disturbing influence and needed to be kept out, declared the governor.

By 1510 there were only 25 000 Caribs still able to work in Hispaniola, and some 250 Africans were sent to work in the gold mines, almost certainly Africans coming from Europe. A report to the King of Spain the next year stated that the work of one African was worth that of four Indians, adding that many of the Africans were used to horses, unlike the Indians. Each imported African was taxed at the rate of two ducats, making the slave trade valuable to the Spanish Crown.

Around 1519, the Spanish Hieronymite friars in Hispaniola had a problem. Most of their Indian slaves had died as a result of the hunt for gold, or from disease, and the few remaining were not enough to do the work required of them. They requested the free entry of African slaves, and this permission was quickly granted, opening the flood gates for all to import Africans into the New World.

At first the number of African slaves remained small, and further south, the first shipment of black slaves to Brazil only happened in 1538, around the time that Brazilian sugar began to reach other markets in Europe. The prospect of riches in the New World was clear to all, but it would take a while for other nations to work out how best to get their share.

HYPOCRAS

Take a gallon of claret or white wine, and put therein four ounces of ginger, an ounce and a half of nutmegs, of cloves one quarter, of sugar four pound; let all this stand together in a pot at least twelve hours, then take it, and put it into a clean bag made for this purpose, so that the wine may come with good leisure from the spices.

Gervase Markham, *The English House-wife*, London, 1615

4
THE ENGLISH
AND THE SUGAR
BUSINESS

Spain and Portugal were the first nations to begin serious exploration, and the first to start laying serious claims to new territories outside Europe. After 1493 they had the authority of the Treaty of Tordesillas to back their claims, and the Portuguese promptly claimed Africa as their own, but the other nations of Europe were unimpressed. In this treaty, Pope Alexander VI specified a line from Pole to Pole passing through a point west of the Cape Verde Islands at about 50 degrees west of Greenwich. This line assigned perpetual ownership, to either Brazil or Spain, of all new lands that were found, anywhere in the world.

In 1529 the Treaty of Zaragoza (or Saragossa) added a further dividing line at about 145 degrees east, completing the division of the globe into Spanish and Portuguese hemispheres, and gave the Philippines to Spain. By the end of the sixteenth century, however, there were Protestants around who cared little for rules made by a Pope, and even Queen Mary could not get her Privy Council to endorse the treaty. Loyalty to nation and profit took precedence over any religious loyalties.

In fact, about the only time that the British, French or Dutch recognised the treaty line was when they argued that European peace treaties had no force beyond the line. In other words, when it suited them the line was there, but at other times it evaporated, and there was 'no peace beyond the line'.

So in Queen Mary's time, the Privy Council gravely forbade African expeditions. They did their duty by saying that much, and the English ships continued to sail. The English knew that even the Catholic French took no notice of the treaty. As far back as 1523, when a French captain called Jean the Florentine had led an attack on a Spanish treasure ship, and King Charles I of Spain had protested over it, King Francis II of France had answered, 'Show me the testament of our father Adam, where all these lands were assigned to your Majesty.'

If there was no problem playing tricks on the Dons under Catholic Queen Mary, there was firm if subtle encouragement once Protestant Queen Elizabeth was on the throne. Thus her Privy Council also told the sailors not to go, and then winked at them as they sailed off. Even if Spanish and Portuguese spies knew all this, there was little they could do, what with the French Calvinists having a colony at Fort Coligny in what is now Brazil, the Portuguese and Dutch squabbling over the Spice Islands and much else as well, not to mention the Dutch, the English, and even the Danes, all poised to join in and take a share of the Spanish–Portuguese cake—and all of those countries got their share of sugar as well.

The German lawyer Paul Hentzner visited London in 1598. He tells us of his visit that:

> . . . [the English] are more polite in eating than the French, devouring
> less bread, but more meat, which they roast in perfection; they put a
> great deal of sugar in their drink; their beds are covered with tapestry,

even those of farmers; they are often molested with the scurvy, said to
have first crept into England with the Norman Conquest . . .

This sugar, though, had dire consequences, Hentzner thought,
as he reported on the English sovereign, Elizabeth I:

> . . . next came the queen in the sixty-fifth year of her age, as we were
> told, very majestic; her face oblong, fair, but wrinkled; her eyes small,
> yet black and pleasant; her nose a little hooked; her lips narrow, and
> her teeth black; (a defect the English seem subject to, from their too
> great use of sugar) . . .

A number among the courtiers he met would have been benefiting from the price sugar commanded, for even as its availability
increased, so did the hunger for sweet tastes, and the courtiers,
as was common, benefited from trade even as they pretended
to despise the traders. They sent ships to the Mediterranean to
trade in sugar, and to Madeira where they bought sweet wines
and sugar, and they also traded in fine sugar from Amsterdam.
By the time Shakespeare posed the apparent puzzle of the
Clown's sugar in 1609, everybody knew about sugar, even if they
had not actually tasted it. Soon, even that would change.

One curiosity, though: the usual editions of Hentzner's
account of his travels feature a drawing of Queen Elizabeth, the
work of one Signor Zuccaro. Given his name, it is hard to avoid
mentioning that, sweetly and wisely, Signor Zuccaro did not
show the Queen's teeth.

PIRATES AND TRADERS

Sir John Hawkyns (to use his own spelling) was a Devon man
like his kinsman, Francis Drake, and he is often called the first
English slave trader. While he did indeed sell a few slaves that he

took from Portuguese ships, his justification would have been that he did not trade regularly in slaves, that he did not take slaves, but that he needed slaves to force a trade with the Spaniards. Much of this was true. A more valid (and more honest) justification might have been that if priests and prelates, rulers and great lords could see no harm in taking and trading slaves, why should he object to making a fortune?

Whatever the case, it was the age of the seadog, and there was just one law at sea: if you saw a foreign ship and thought you could capture it, you did—and if you thought it could capture you, you fled. It was a seadog-eat-seadog world—that was something Hawkyns knew well, and so did all the Devon men.

Thomas Wyndham was very much the genuine seadog. He had sailed with Hawkyns' father and later mounted his own expeditions. In 1552 Wyndham was master of the *Lion*, when he was forced to land in the Canaries, on a small island between Fuerteventura and Lanzarote, to mend a leak below the waterline. His crew took 70 chests of sugar ashore to lighten the ship, but these chests were seen by islanders who claimed they came from a ship that had just left port, which led to Wyndham being accused of piracy.

The matter was sorted out easily enough when Wyndham's men captured the governor, a man he described as 'a very aged gentleman of seventy'; being thus well placed to negotiate, they made good their departure. A year later Wyndham died on the way home after he had sailed to the African coast with another ship, seeking trade in gold and pepper, but he had shown the way to work. You can negotiate with the Dons, said the Devon men, but you need to get them over a barrel first, if you want the best of the bargain. That was the norm for trade in those times, and if Wyndham had in truth taken the chests of sugar, well, that would not have been unusual either.

In 1562 John Hawkyns was 30 years old, and he was ready to go trading. He sailed for Africa, planning to go after things like gold dust and ivory, materials which had a ready home market. He sailed for Africa in three ships, with the financial backing of the treasurer of the navy (it helped some that this official was also his father-in-law), two city magistrates, the Lord Mayor of London, a future Lord Mayor and, most importantly, Queen Elizabeth herself. They captured 300 slaves, mainly by taking them from Portuguese ships headed for the Cape Verde Islands and, having annoyed the Portuguese, set out to tweak a few Spanish beards.

Arriving at Hispaniola, Hawkyns claimed he needed to careen his ships and that he could only pay for this by selling some of the slaves. Then, having opened up the trade to raise careening money, he opened it up a little more and so managed to return to England with a clear profit. That, at least, was the story the two sides told—it is likely that the Spanish colonists, chafing under trading restrictions imposed by the home government, were happy to play along with a neat cover story.

On his third voyage, Hawkyns sailed with six ships, two of them belonging to the Queen herself. At the Spanish colonial port of Rio de la Hacha, the English fleet fired off a few cannon and took the town. Once they were ashore and safely in charge, two slaves, one a mulatto, the other a Negro, revealed where some Spanish treasure was hidden in exchange for help in gaining their freedom.

Hawkyns claimed afterwards that because he was an honest trader he took only 4000 pesos from the treasure for each of the slaves he left in the town, and returned the rest. Then, because the loyalties of race and class counted for more than the loyalties of nationality or religion, he handed over the slaves who had so treacherously revealed where the treasure was hidden. The

Spaniards, equally keen to respect this assistance from a gallant and honourable adversary, and sensitive to distinctions of either race or guilt, promptly quartered the Negro, and hanged the mulatto, both for treason.

Once again, the story may be open to some doubt, since the Spaniards claimed the slaves they were forced to buy were old and feeble, sickly and dying. Perhaps the execution of the traitorous slaves was a fiction added to the tale to make it sound better, or perhaps they really were done to death so that the guilty parties would not meet a similar fate.

Hawkyns' modern English apologists argue that the slaves he sold were used to extort money from the Spanish and to stimulate trading—so he was not so much trading in slaves as using them so that he could trade. The end result, though, was that his third voyage brought the first West Indies sugar into England, and Francis Drake had gained valuable experience in dealing harshly with the Spanish. You needed a fierce resolve, cold steel, iron cannonballs and plenty of lead musketballs to trade on an even footing with the Dons.

Curiously, lead is a recurring theme in the story of sugar. Alexander VI, the Pope who approved the Treaty of Tordesillas, father of Cesar and Lucrezia Borgia (among others), bribed and poisoned his way to the papal throne. He probably died of a fever, though at the time there were plenty willing to believe that he had accidentally drunk poisoned wine set aside for Cardinal Corneto, and so suffered a just fate. That, at least, is the story that Alexandre Dumas told in one of his novels.

The poison the Borgias used was probably lead acetate, a soluble lead salt, known from its sweet taste as 'sugar of lead'. It is likely that Corneto's wine, if it was his, was laced with this. But whether or not Pope Alexander VI died in this ironic (leadic?) way, we will probably never know.

There was another link between lead and sugar, apart from the use of lead pipes to carry cane juice from the mill to the boiler-house. Around AD 1000, lead acetate was used in Egypt as a defecant, an agent to clean the heated cane juice, but using lead like this was soon banned. After that time, suspect syrups were exposed near a latrine where the action of hydrogen sulfide coming from the cesspit would produce a tell-tale black precipitate of lead sulfide in the syrup.

In 1847 a British patent proposed the use of lead salts in the preparation of sugar, an idea which alarmed so many people that Earl Grey felt the need to send a circular to all British colonial governors, warning them against allowing it.

THE INDENTURED SERVANTS

Given the piratical habits of all sides, it is hardly surprising that when *William and John*, the first British ship to reach Barbados, arrived there in 1627, it carried 80 English settlers, and also half a dozen Negroes plundered from a Portuguese vessel 'met' on the way. On the return voyage, the crew captured a Portuguese ship with a cargo of sugar. This cargo was sold for £9600, which went to benefit the colonists. While the colony might thus seem to have begun on a combination of black slaves and sugar, it really began with indentured white labour, and with crops other than sugar.

A year after the British had first settled on Barbados, Henry Winthrop reported a mere '50 slaves of Indeynes and Blacks'— and that included the blacks collected on the way to the island. Between 1628 and 1803 the island imported 350 000 slaves, of whom 100 000 were women, but when the last of the slaves were freed in 1834 they were just 66 000 in number. Many of those would have been born after imported slaves stopped arriving, for when slaves could no longer be shipped in, breeding was encouraged. For a comparison, in the United States, between 1803 when the importing of slaves was officially banned, and 1865, the slave population increased tenfold due to internal population increase.

The white indentured servants of the seventeenth-century colonies were seen as people excess to the needs of the home nations. As early as 1610, Governor Dale of Virginia pointed out that the Spanish had greatly added to the (white) populations of their American colonies by sending out their poor, their rogues, their vagrants and their convicts. In 1629 Henry Winthrop realised that he needed 'every yere sume twenty three servants' to work his tobacco plantation in Barbados. While these were not always available, the English Civil War began to provide shipments of prisoners in 1642. Soon a group of kidnappers known as the Spirits became active, 'crimping' or kidnapping people who found themselves carried to Barbados, where the ships' captains would, in effect, sell them into slavery.

Many of these servants were seen as troublemakers in their new homes. The Scots servants were rated more highly than the English; the Irish servants were rated so poorly that the Barbados Assembly enacted a law in 1644 against any increase in their numbers. Because Barbados had to take what it could get, however, some 20 per cent of servants were still Irish in 1660.

In the meantime, yellow fever had come to the New World with the African slave ships. Spread by the mosquito *Aedes aegypti*, it had to wait until a fast trip carried a single generation of mosquitoes across the Atlantic from Africa in the open barrels of foul drinking water. Once they reached the islands and South America, the insects dispersed and passed the disease on. To European populations, this tropical disease was a serious threat. Those who survived never got it again, so an immune population of survivors eventually developed, but new arrivals were always at risk.

A yellow fever epidemic on Barbados around 1647 killed an estimated 6000 whites, many of them indentured servants. This increased the demand for black slaves in the 1650s, but for the moment the wars in Britain kept up the supply of indentured servants. As the number of black slaves increased, cheap white labour was needed to keep control. An Act of 1652 provided a solution, allowing two justices of the peace to

> . . . from tyme to tyme by warrant . . . cause to be apprehended, seized on and detained all and every person or persons that shall be found begging and vagrant . . . to be conveyed into the port of London, or unto any other port . . . from which such person or persons may be shipped . . . into any forraign collonie or plantation . . .

In other words, the magistrates, who represented the wealthy of a borough or parish, could ensure that the poor, who otherwise would be a cost to the parish, were rapidly and permanently removed. Note the use here of 'plantation' as a synonym for 'colony'. Up until about 1650 it was people who were thought of as planted, and settlements of English families in Wales or Scotland were also 'plantations'. It was only later that plantation came to mean a single (generally monocultural) farm. When first settled, Barbados was a plantation of people; it became a sugar plantation when sugar cane was introduced, and then a sugar colony filled with sugar plantations.

THE WHITE SLAVES

The indentured servants who survived the yellow fever probably saw themselves as fortunate, especially those who had gone to the

NORTH ATLANTIC OCEAN

A topographicall Description and
Admeasurement of the YLAND of
BARBADOS in the West INDYAES
(Redrawn from Richard Ligon's map of 1657.)

Key:
Sugar plantation
0   2 miles
0   2 kilometres

For the most part, the plantations of Barbados were clustered near the coast.

islands of their own accord and were able to hire themselves out. The servants who had been 'Barbadoed' by a court on a trumped-up charge, or stolen away from their families by the Spirits and sold into virtual slavery in the islands, might have accounted themselves fortunate to get out of plague-ridden London, but many died of island diseases instead. For some this would have been a happy release. It was the island diseases that kept servants in short supply, so that judges in England and Ireland would happily find prisoners guilty and send them to the West Indies, or 'Barbadoe' them, as the saying went. It was why the Spirits were able to operate in the various ports; because there was such a shortage of servants they could pay well for people to turn a blind eye, knowing they would be well paid for all they captured.

The Spirits got up to all sorts of tricks, the same ones the crimps of that time played on sailors, using knock-out potions or getting people drunk, and sending the victims to sea with forged papers showing they were indentured. This meant the victim would spend seven years working for a master who generally cared little for the welfare of a servant who would be lost to him at the end of that time.

Of course, not all the servants could get away from their indentures. Under a code passed in 1661, setting out the rights of master and servant, the servants, just like slaves, were forbidden to engage in commerce, and any of a large number of petty offences could lead to a year being added to the period of servitude. Passes were required for servants to be off their master's property, and the dishonest master had plenty of opportunities to goad servants into punishable actions so their periods of servitude could be extended.

Being indentured was so bad that when Generals Penn and Venables were in Barbados in 1654 to outfit for their attack on Hispaniola and Jamaica, many servants ran away to the ships—

while those on the ships, knowing what life at sea was like, fled ashore. In the end, Cromwell's great Western Design to undo Catholic Spain meant 2000 servants from Barbados perished on Jamaica, so in February 1656 Cromwell sent troops into London to find 1200 women of 'loose life' and send them to Barbados. Within days another 400 were sent off.

The planters always complained about the servants. Even though Scotland was under the same king, it was still a different country, and the planters asked in 1667 to be allowed free trade in servants with Scotland, and to transport 1000 to 2000 English servants to the colony. Still nobody wanted the Irish, because it was believed that, being Roman Catholic, they were likely to help the French or the Spanish if they could. The proportion of servants to slaves was a bigger worry, however, so the planters still took what servants they could get.

By 1680 there were only 3000 indentured servants on Barbados, down from 13 000 in the 1650s. By that time there were so many slaves that the masters used all sorts of legal tricks to hold their indentured servants, but those who were out of indenture could pick and choose where and how they worked.

When the Monmouth rebellion broke out in 1685, the planters got a break. The bastard Duke of Monmouth tried to seize the British throne but failed, and his followers were either put to death or, more often, Barbadoed. The Spirits had a fine old time, snatching extra bodies and sending them off with papers showing their victims as convicted rebels.

Most of the former servants had skills that were badly needed, and they could set their own price. Many of the advances in sugar preparation (like the Jamaica Train discussed in Chapter 6) must have come from servants who had reached sugar master status. They had got their training thanks to the Spanish Inquisition and the way the inquisitors had treated the Portuguese Jews, who had been happy and safe in Portugal until a few years before the Great Armada, when Spain took over Portugal and the Inquisition moved in on the Jews.

Most of the Jews in Portugal had fled Spain and its Inquisition a generation or two earlier, and now they shifted again, to Holland, where they were welcomed for their skills. Some of them moved to South America when the Dutch took over Pernambuco in the north of Brazil. The Jews, seen as an under-class, managed the daily operations of the plantations, and more importantly, the mills. When the Dutch were forced out of Pernambuco, many of the Jews went with them to Amsterdam, but others went to Barbados, where they provided the skills base that the English sugar planters desperately needed.

THE ROYALIST REFUGEE

During the turmoil of the English Civil War of the 1640s, the royalist Richard Ligon felt it would be safer to be out of England than to stay there. So he took himself off to the peace and calm of the plantation of Barbados in 1647, not returning home until 1650, by which time life in England was a little more stable.

He spent a pleasant enough three years, learning the art of sugar making among other things, and set down what he saw in *A True & Exact History of the Island of Barbados*. Because he was there long enough to observe closely, but not long enough to become part of the community, Ligon's account gives us the truth, hopefully unvarnished by any desire to censor the facts. For example, he explains that:

> The slaves and their posterity, being subject to their Masters for ever, are preserv'd and kept with greater care than the servants, who are

theirs but for five years, according to the law of the island. So that for the time, the servants have the worser lives, for they are put to very hard labour, ill lodging, and their dyet very slight. Most of them are Irish and a sullen bunch, but that may be on account of the treatment meted out to them, I know not. The usage of the Servants, is much as the Master is, merciful or cruel. Those that are merciful, treat their Servants well, but if the Masters be cruel, the Servants have very wearisome lives.

Before his time in the island, he tells us,

... the first people in Barbados, made tryal first of tobacco, cotton, indigo, and only then turned to sugar canes. The planters made tryal of them and finding them to grow, they planted more and more, as they grew and multiplyed on the place, till they had such a considerable number, as they were worth the while to set up a very small *Ingenio*, as we call the place where crushing and boiling down to make the sugar takes place.

At the *Ingenio*, he reports, the cut cane is placed on a platform called a *Barbycu*, a raised stand with a double rail to stop the cane falling out, about 9 metres long and 3 metres wide (in spite of the size, this was a close relation to our modern barbecue).

Then there is a set of three rollers, with perhaps five horses or oxen driving the middle roller, 'which is cog'd to the other two, at both ends':

A *Negre* puts in the canes of one side, and the rollers draw them through to the other side, where another *Negre* stands and receives them and returns them back on the other side of the middle roller which draws the other way. So that having past twice through, that is forth and back, it is conceived that all the juice is prest out; yet the Spaniards have a press, after both the former grindings, to press out the remainder of the liquor ...

But that, he explains in a bluff patriotic manner, is because the Spaniards' cane is poorer. The crushed cane is set aside, some six-score paces away, and the juice:

> ... runs under ground in a Pipe or gutter of lead, cover'd over close, which pipe or gutter, carries it into the Cistern, which is fixt neer the staires, as you go down from the Mill-house to the boyling house. But it must not remain in the Cisterne above one day, lest it grow sowr; from thence it is to passe through a gutter, (fixt to the wall) to the Clarifying Copper ... As the skumme rises, it is conveyed away, as also the skumme from the second Copper, both which skimmings, are not esteem'd worth the labour of stilling; because the skum is dirtie and gross: But the skimmings of the other three Coppers, are conveyed down to the Still-house, there to remain in the Cisterns, till it be a little sowr, for till then it will not come over the helme and make good rum.

> ... there is thrown into the four last Coppers, a liquor made of water and ashes which they call Temper, without which, the Sugar would continue a clammy substance and never kerne. Once the sugar master has determined that the sugar in the last copper is ready, two teaspoonfuls of Sallet Oyle [salad oil], such as we put on raw vegetables to make a sallet, are added and then the syrup is ladled out. Above all, it is important to throw in some cold water, in order that the last of the syrup should not burn, for the copper is fixed in place over an open fire, and as soon as the copper is empty, syrup from the penultimate copper must be added.

> ... And so the work goes on, from Munday morning at one a clock, till Saturday night, (at which time the fires in the Furnaces are put out) all houres of the day and night, with fresh supplies of men, Horses and Cattle. The liquor being come to such a coolness, as it is fit to be put in the Pots, they bring them neer the Cooler, and stopping first the sharp end of the Pot (which is the bottom) with Plantine leaves, (and the passage there no bigger than a man's finger will go in at) they fill the Pot and set it between the stantions in the filling room,

where it staies till it be thorough cold, which will be in two days and two nights; and then if the Sugar be good, to be removed into the Cureing house, but first the stopples are to be pulled out of the bottom of the pots, that the Molosses may vent itself at that hole.

. . . The Molosses, in a well-run Ingenio, is converted into Peneles, a kind of Sugar somewhat inferiour to the Muscovado. . . . And this is the whole process of making the Muscovado Sugar, whereof some is better, and some worse, as the Canes are; for, ill Canes can never make good Sugar.

I call those ill, that are gathered either before or after the time of such ripeness, or are eaten by Rats and so consequently rotten, or pulled down by the vines men call Withes, or lodged by foule weather and ill winds, either or which, will serve to spoil such Sugar as is made of them.

A major improvement in the lot of the planters came when rum became part of the sugar industry. It made marginal operations profitable, loss-making plantations became profitable, and planters still losing money found a new comfort. Almost nothing was too poor to go into the fermentation vats, other than the first couple of skimmings from the coppers. Richard Ligon is once again one of our best witnesses:

After it has remained in the Cisterns . . . till it be a little soure, (for till then, the Spirits will not rise in the Still) the first Spirit that comes off, is a small Liquor, which we call low-wines, which Liquor we put into the Still, and draw it off again; and of that comes so strong a Spirit, as a candle being brought to a near distance, to the bung of a Hogshead or But, where it is kept, the Spirits will flie to it, and . . . set all afire.

This volatility of the rum made it quite risky, and Ligon describes how they 'lost an excellent Negro' to a rum explosion when a candle was used for illumination while a jar of spirit was being added to a butt of rum:

. . . the Spirit being stirr'd by that motion, flew out, and got hold of the flame of the Candle, and so set all on fire and burnt the poor Negro to death, who was an excellent servant. And if he had in the instant of firing, clapt his hand on the bung, all had been saved; but he that knew not that cure, lost the whole vessel of Spirits, and his life to boot . . .

This drink, though it had the ill hap to kill one Negro, yet it has had the vertue to cure many; for when they are ill, with taking cold, (which they often are) . . . they complain to the Apothecary of the Plantation, which we call the Doctor, and he gives to every one a dram cup of this Spirit, and that is a present cure. And as this drink is of great use, to cure and refresh the poor Negroes, whom we ought to have a special care of, by the labour of whose hands, our profit is brought in; so it is helpful to our Christian servants too . . .

The distinction they made between their servants may seem an odd one, but Ligon explains even this:

Once I encountered a slave who wished to be a Christian, but on interceding with the slave's master, I was told that the people of the Island were governed by the Lawes of *England*, and by those Lawes, we could not make a Christian a Slave. I told him, my request was far different from that, for I desired him to make a Slave a Christian. His answer was, That it was true, there was a great difference in that: But, being once a Christian, he could no more account him a Slave, and so lose the hold they had of them as Slaves, by making them Christians; and by that means should open such a gap, as all the Planters in the Island would curse him.

To read the testimony of the planters, nothing was ever easy for them. That is one point at which Ligon was in complete agreement with later writers with the mindset of the plantation owner.

DERBY'S DOSE

The planter could make a good profit, but there was always the risk of bad weather, insurrection or war, not to mention death from disease (or taxes from a home government). The planter had to buy, clear and plant the land, buy Guinea grass for the animals, set up gardens for the slaves, and general working and living space. Purchases included tools, nails, hoops and staves for barrels, lime, cooking pots, building material, food for the slaves, equipment for the mill and boiling house, and then there were the skilled staff: even if these were slaves, they could still command extra allowances—and many of them were free men, former indentured servants now out of their indentures.

The overseer, distiller, carpenter, drivers and wainmen, cooper, foreman sawyer, fireman, watchman, field-children's nurse, potter and 'black doctor' had all to be paid, as well as domestic servants. But above all, the slaves had to be fed, and while bought food was expensive, the food crops perversely needed the most cultivation just when the sugar needed harvesting!

The slaves were fed well enough at times, though for the most part planters tried to keep costs down by using local resources. The areas between cane plots could be planted with food crops, including such crops as yams, eddoes and bananas, brought from Africa by the slave ships. William Bligh's ill-fated breadfruit was one of the few failures; the slaves did not like the taste, and it was only well into the nineteenth century that people in the Caribbean began to eat it.

Hard physical labour requires protein, and sweaty work requires salt. It did not take long for the canny cod fishermen of New England to identify a new and not particularly fussy market. Their rejects, the badly split fish and fish with too much salt or not enough, could all be disposed of as 'West India cure', destined to feed the slaves. During the eighteenth century, in times of unrestricted trade, on average a ship would leave Boston every day for the West Indies, laden with reject fish. Around 1650, Richard Ligon saw that fish could be found closer to home, and he wrote in his *Exact History*:

> As for the *Indians*, we have but few, and those fetcht from other Countries; some from the neighbouring Islands, some from the Main, which we make slaves: the women who are better vers'd in ordering the Cassavie and making bread, than the *Negroes*, we imploy for that purpose and also for making Mobbie; the men we use for footmen and killing of fish, which they are good at; with their own bowes and arrows they will go out; and in a dayes time, kill as much fish as will serve a family of a dozen persons, two or three dayes if you can keep the fish so long.

Other foods for the slaves varied from island to island. Jamaica had more free land than Barbados, enabling the slaves there to tend gardens where they grew food. Barbados was necessarily more dependent on outside sources, importing maize and rice from America and horse beans from Britain. Reliable figures are hard to come by, but one record exists of newly purchased slaves with no planted ground getting one fish and either nine plantains, two pints of rice or three pints of maize, each day. The food rations, it would seem, were minimal and monotonous, and might have accounted for the short working lives of most slaves.

During the sugar harvest there was cane to chew and syrup to drink, but by the end of the harvest the provision grounds were least productive. Thomas Thistlewood, an overseer in Jamaica in the middle of the eighteenth century, recorded signs of poor nutrition among the slaves in August and September, over a number of years. His diary for 25 May 1756, as quoted by Ward, reveals that a slave called Derby was caught eating the young canes—a definite offence, since it meant a reduced crop later on:

'Derby catched by Port Royal eating canes. Had him well flogged and pickled, then made Hector shit in his mouth.' This treatment, referred to thereafter as 'Derby's Dose', did not seem to deter the offender, who appears in Thistlewood's diary again in August:

> Last night Derby attempting to steal corn out of Long Pond corn pieces, was catched by the watchman, and resisting, received many great wounds with a mascheat [machete], in the head etc. Particularly his right ear, cheek and jaw, almost cut off.

There is no record of what happened to Derby after that. It seems unlikely he survived, though. The excerpts from Thistlewood's journal quoted by Ward clearly reveal his care and consideration for those under his charge— when they weren't stealing food, that is. All sorts of odd punishments, including lockable masks of tinplate, were used to stop slaves eating the young cane. Wearing the mask was probably preferable to what happened to Derby. In some areas, the masks were also worn by kitchen slaves to stop them tasting the food they were preparing.

Forced to labour in the sugar mills from sunrise to late at night, it was inevitable that weary slaves would lose concentration and risk injury. The greatest danger came at the height of the season, when they were toiling away close to huge pans of sticky, scalding sugar juice, working for up to eighteen hours a day in fierce heat in the rush to deal with the huge masses of ripe cane that came in. Outside, the slaves who fed the cane into the rollers worked in cooler conditions, but they were just as much at risk of injury through being trapped by the rollers.

The three-roller mill was standard, and while some early ones were powered by humans, most were under animal, wind, water or steam power and slower to react to a human scream. A hatchet or cutlass was kept in a convenient place, ready to chop off the arm of any slave who was trapped—in order to save the slave's life. It made better economic sense to keep alive a slave with one arm, because that slave could still act as a watchman, clear blocked drains or guide the animals that did the heavy haulage.

While there is probably a degree of exaggeration in the tales the emancipists told later of slavery, it would be unwise to assume that the life of a slave was a pleasant one. The way slaves took advantage of unrest made this very clear—in fact, slaves were one reason not to fight wars in the Caribbean.

The territorial claims made by Spain and Portugal under the Treaty of Tordesillas could not be defended against the combined forces of the English, Dutch and French, and in the end Spain was forced during the seventeenth century to accept the presence of other powers in the Caribbean islands, just as Portugal had to accept the British and French in India and Africa, and the Dutch in the East Indies. In one case, the Spaniards shared the island of Hispaniola with the French.

Aside from anything else, whatever armed forces the Europeans had in the West Indies were needed there to maintain order in their own colonies. When the home nations went to war, extra forces would be sent in to attack and pillage the colonies and the shipping of the enemy, even though this caused unrest among the slaves.

TO DRIE APRICOCKS, PEACHES, PIPPINS OR PEARPLUMS

Take your apricocks or pearplums, & let them boile one walme in as much clarified sugar as will cover them, so let them lie infused in an earthen pan three days, then take out your fruits, & boile your syrupe againe, when you have thus used them three times then put half a pound of drie sugar into your syrupe, & so let it boile till it comes to a very thick syrup, wherein let your fruits boile leysurelie 3 or 4 walmes, then take them foorth of the syrup, then plant them on a lettice of rods or wyer, & so put them into yor stewe, & every second day turne them & when they be through dry you may box them & keep them all the year; before you set them to drying you must wash them in a little warme water, when they are half drie you must dust a little sugar upon them throw a fine Lawne.

Elinor Fettiplace's Receipt Book, 1604

5
FIGHTING OVER SUGAR

Bryan Edwards, a planter and early historian of the West Indies, explained war in his neighbourhood like this:

> Whenever the nations of Europe are engaged, from whatever cause, in war with each other, these unhappy countries are constantly made the theatre of its operations. Thither the combatants repair, as to the arena, to decide their differences.

According to Edwards, this was because the combatants who survived could make themselves rich. In the late eighteenth century, foreign navies plundered British merchant ships and kept the profits while Britain's navy made a treasure trove of the foreign trading vessels. But did the navies compensate the planters for their losses? Indeed they did not, the planters complained. The arena for their grudge matches was inevitably the lucrative Caribbean, but paying compensation to the planters would have eaten into their profits.

The warring navies chose the Caribbean, far from their home waters, for the rich cargoes carried in the area, and because of the way that prize money works in times of war, especially

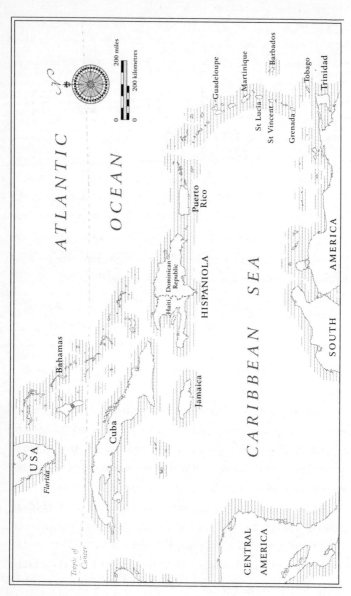

The West Indies.

benefiting frigate captains whose ships were large enough to sail independently, and fast enough to run down almost any ship on the ocean. Edwards conceded that sometimes the British planters would gain, since Britain usually held the upper hand in privateering and blockading. This meant the French sugar trade was often badly affected, allowing English sugar interests a greater slice of the European market. At the same time, Royal Navy ships provided a ready market for rum, but the planters were not happy—it was not in their nature to be happy.

Prize money was paid for all ships and cargoes captured. It was divided in a complex manner, with larger sums going to the more senior officers, and many captains—if they survived long enough—became landed gentry in their later years. Frigates did best of all, because if a capture was out of the sight of the commanding admiral, the admiral's portion was also divided among the officers and crew.

Sometimes the naval officers were a bit greedy. Tradition has it that Josias Rogers, captain of the *Quebec,* was so impressed by the sight of a bullion-laden Spanish treasure ship, brought into Portsmouth during the Seven Years' War, that he determined to enter the navy and have a share in such riches. He did quite well from the War of American Independence, and settled on an estate in Hampshire, but when his banker failed and he lost half his fortune, Rogers just went back to sea to get some more. In the first five weeks of 1794 he took nine prizes, and estimated that his share of the proceeds would be £10 000.

The Royal Navy had taken more than 300 merchant ships in 1794, mainly American neutrals, in this legalised form of plunder. The prize courts later rejected half the claims on the ground that these neutral ships were sailing between neutral ports and not subject to seizure, but Captain Rogers and his crew still gained from three of their nine prizes. He later spent £3000 in contesting the lost cases, but Rogers did not enjoy his restored wealth for long, however—he saw both his younger brother and nephew die of yellow fever before he succumbed to the same disease in 1795. There were rich pickings for those who survived, but many more lost their lives to disease.

The naval physician, Sir Gilbert Blane, found that in one year alone, 1779, England's West Indies fleet lost an eighth of its 12 019 seamen to disease—a total of 1518 dead, with another 350 'rendered unserviceable'. In 1794, the then Vice-Admiral Jervis' West Indies squadron lost about a fifth of its men to disease in just six months. The 89 000 soldiers of all ranks serving in the West Indies between 1793 and 1801 suffered 45 000 deaths, 14 000 discharged and 3000 desertions. Small wonder that British troops being sent to the West Indies were usually sent first to the Isle of Wight or Spike Island in the Cove of Cork, to prevent them deserting *en masse*. German and French mercenary units particularly objected to being sent to what they saw as certain death, and either deserted or mutinied at the prospect. While soldiers could also earn prize money, there was generally less to be had on land than on sea, and a much better chance of falling to disease.

This was why the army and the navy had different views of war in the islands. On land, yellow fever was almost a certainty, and too many of the soldiers died of disease, trying to win from France sugar islands Britain did not need. Henry Addington, arguably Britain's worst Prime Minister, was not exaggerating when he later told the House of Commons that the West Indies had destroyed the British army. Still, the navy was happy, because of the rich pickings, while the government could only see the French losing sugar and sugar income.

The rich pickings included vast quantities of molasses, rum and raw sugar. But while sugar was undoubtedly the most valuable booty, there were other riches to be had from the Caribbean—coffee, cocoa, cotton, indigo and ginger. While most of these products came also from other places, and Britain was never much of a market for coffee (because the East India Company's tea was more popular), these cargoes were all of value in the European markets. That raises a question: why did people grow so many of these things in one place, concentrating the riches and increasing the risks?

Part of the answer lies in the wind patterns of the Atlantic, because all the produce had to be carried to distant markets in sailing ships. Part lies in the climate of the islands, and part lies in the fact that the plantations were close to the sea, allowing ready transport to the ocean-going ships that carried the cargo away. Wind patterns established the two main triangular trades: dried cod from North America to Africa, slaves from Africa to

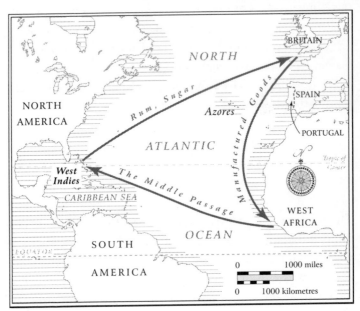

The triangular trade.

the Caribbean and molasses or rum from the Caribbean to New England; and cheap manufactured goods from England, mainly textiles from Lancashire and hardware and toys from Birmingham, to Africa, where these were converted into gold dust, ivory and pepper, and slaves for the Caribbean leg, where the proceeds were used to buy sugar, molasses and rum for the homeward journey.

New England ships sometimes sailed an even tighter loop, taking rum to Africa, slaves to the West Indies, and then molasses back to New England to repeat the cycle when the molasses was turned into rum. French, Dutch, Danish and Portuguese ships did similar circuits, with variations—for example, cod to the Canary Islands, and wine from the Canaries to Africa, where it was exchanged for slaves. All of these trades depended on the fact that the quickest way north from Africa involved travelling to the Caribbean first, whether the voyage was to England or New England. Yet through all of this, with so many people benefiting from the trade, the planters were the only ones blamed for all the evil people said was being done.

The planters survived the wars—unlike the soldiers. Well, some of the planters did—some of them fell victim to their own slaves.

FREEDOM FIGHTERS

The white people on the islands believed that coloured nurses could kill young children without trace by using a scarf pin pushed into the head, and whether this was true or not, it was enough to feed the fear. Macandal was a slave who probably lost his arm in a roller accident, but rather than work at clearing drains he ran away into the hills of Saint Domingue. The one-armed runaway started a campaign of poisoning in 1750, slipping quietly into plantations and providing poison to unsuspected accomplices, until he was apprehended and burned alive in 1758.

There were others, less well known but equally effective, because the planters used slaves in their houses as servants, as cooks, as minders of their children, and as mistresses. In a time when the whites showed no mercy to the slaves, the slaves showed no mercy in return.

The planters were right to fear their slaves. The first slave revolt occurred on São Tomé in 1517, and others over the next hundred years caused much damage to that island's economy. In 1522, black slaves on Hispaniola rose up in the first revolt in the New World, but the first sugar-colony black revolt did not take place until 1656, when two slaves from Angola, Jean and Pedro Leblanc, led a revolt on the French island of Guadeloupe.

There might have been a successful uprising in the British colony of Antigua in 1736 but that one conspirator was arrested for a minor offence. Thinking that he was under arrest for helping plan the coming revolt, he told all. In the end, five slaves were broken on the wheel, five were gibbeted, and another 77 burnt alive. While the loss of human life was no great matter in those days, the loss of property was, and the waste of so many workers shows the fear the white planters were in. And they knew there were always others to try once more.

The Bastille was stormed in Paris in July 1789, but there was no Liberty, Fraternity and Equality for the slaves in the French colonies. On 20 March 1790 the National Assembly stated that the declaration of the rights of man was not to apply to the colonies. Saint Domingue at that time had three classes of citizens: 30 000 white planters and officials, 24 000 *sangs mêlés* (people of colour) and half a million slaves, along with a few freed blacks. Until 1777, the *sangs mêlés* had been allowed to go to France for their education and, unlike the slaves, they were well off, with hopes and aspirations.

The French Revolution had divided the colony's white population, with its leaders remaining royalist while the masses took up the revolutionary cause. It was against this background that the slaves sought their freedom. The three Ogé brothers, Jacques, Victor and Vincent, together with a *sang mêlé* called Chavane, failed in an insurrection. Jacques and Chavane were broken on the wheel and Vincent was hanged. Victor escaped and was never seen again.

Back in France, a nervous National Assembly passed a decree allowing people of colour the right to sit in parochial and colonial assemblies. When the planters in Saint Domingue ignored this, their slaves rose up in some areas. The *sangs mêlés* rebelled in the south, but with no support from the slaves there they reached a peace a short time later, which included an amnesty and acceptance of the decree giving them equal rights. Soon, though, the decree was repealed.

Three commissioners arrived from France to take control in January 1792. They brought 6000 troops with them, but the commissioners were advanced revolutionaries. Finding themselves opposed by the planters, who were largely royalists and supporters of the *ancien régime*, the commissioners called on the slaves for support. This was just the opportunity the slaves needed, and those whites who did not escape to the ships in the harbour were killed. When war broke out between France and Britain in August 1793, the British invaded Saint Domingue, but in the end the slaves won—with the help of disease. Disease won most of the wars when people came to fight in the West Indies, and there were many wars fought over the sugar islands.

As we have seen, one sugar island was shared. The French founded their colony of Saint Domingue on the western side of

Hispaniola in 1697, while the Spanish claimed the eastern side of the island as San Domingo. By the end of the eighteenth century, the French colony was the most profitable sugar producer in the New World, and it was sacrificed by mismanagement and misunderstanding. The freedom the revolutionary Jacobins gave themselves in France was not extended to the slaves of Saint Domingue, and the planters found themselves facing a revolution led by a very clever general.

Toussaint L'Ouverture was the son of an African chieftain, which gave him authority amongst his fellow slaves, and he had been given some education. More importantly, he was a leader, and in 1793 he put his leadership to the test, calling on his fellow slaves to join with him in bringing liberty and equality to Saint Domingue. He formed an alliance with the Spaniards on San Domingo until 1794, when the French ratified an act which set the slaves free. He then turned upon his Spanish allies and expelled them. As the whole island was now effectively French, the English decided to attack and make it English, and the freed slaves looked like becoming unfree once more.

The English got a toehold in the south, but an outbreak of yellow fever defeated them. They withdrew in 1798 when Toussaint promised that all remaining colonists would be spared, and that he would not invade Jamaica. By now the people of Saint Domingue were calling themselves the people of Haiti, and looking to free other slaves.

This would have been disastrous for the English, who had quite recently made Jamaica more secure by transporting the majority of the maroons, the escaped slaves of Jamaica, first to Nova Scotia and then to Sierra Leone. The English were terrified that the Haitians might inspire the Jamaican slaves to revolt. After all, they knew Toussaint to be a dangerous man, from a letter he wrote to the Directory, France's ruling body before Napoleon seized power, seen here in James' translation:

> We know that they seek to impose some of them on you by illusory and specious promises, in order to see renewed in this colony its former scenes of horror. Already perfidious emissaries have stepped in among us to ferment the destructive leaven prepared by the hands of liberticides. But they will not succeed. I swear it by all that liberty holds most sacred. My attachment to France, my knowledge of the blacks, make it my duty not to leave you ignorant either of the crimes which they meditate or the oath that we renew, to bury ourselves under the ruins of a country revived by liberty rather than suffer the return of slavery. . . . But if, to re-establish slavery in San Domingo [the Decree of 16 Pluviôse were revoked], then I declare to you it would be to attempt the impossible: we have known how to face dangers to obtain our liberty, we shall know how to brave death to maintain it.

Toussaint took much of the credit for the withdrawal of the English, and the French made him a general, with the titles of Lieutenant-Governor and Commander-in-Chief of the Haitian forces. The standard wisdom, though, is that Toussaint did not make the revolution, it was the revolution that made Toussaint.

Toussaint declared himself in 1801 to be Governor-General for life, showing little respect for Haiti's official owner, Napoleon Bonaparte. French honour had to be satisfied—and French sugar and sugar incomes had to be restored. The French attacked, led by Napoleon's brother-in-law Leclerc. The military response to the heat of the tropics was to confine the soldiers to their barracks during the hottest hours of the day—where the mosquito lurked, ready to spread yellow fever, and where troops all too easily succumbed to the temptation offered by cheap rum. If that was not bad enough, in an act of absolute stupidity the French reinstated slavery, which united all of Haiti against them.

In spite of the skyrocketing death rate from disease, the French managed to capture Toussaint by treachery after an armistice, and he was hauled to a prison high in the French Alps where he died soon after. His unhappy end was reflected in a sonnet by Wordsworth:

To Toussaint L'ouverture

Toussaint, the most unhappy man of men!
Whether the whistling Rustic tend his plough
Within thy hearing, or thy head be now
Pillowed in some deep dungeon's earless den;
O miserable Chieftain! where and when
Wilt thou find patience? Yet die not; do thou
Wear rather in thy bonds a cheerful brow:
Though fallen thyself, never to rise again,
Live, and take comfort. Thou hast left behind
Powers that will work for thee; air, earth, and skies;
There's not a breathing of the common wind
That will forget thee; thou hast great allies;
Thy friends are exultations, agonies,
And love, and man's unconquerable mind.

The French might have taken Toussaint, but they had lost 50 000 men, and in the end they lost their most important sugar colony for good. Leclerc died of yellow fever in October 1802, and his replacement, General Rochambeau, was equally unable to overcome the destruction of his army by disease, and he capitulated in November 1803.

Napoleon decided to sell his American colonies to the United States for $15 million. The grand plan he had conceived—to not only retain the sugar wealth of Saint Domingue, but to build a huge French empire in the Mississippi Valley to make up for the loss of Canada during the Seven Years' War—had come to naught.

With the gain of this new territory—the Louisiana Purchase—the United States was able send the explorers Lewis and Clark to reach the west coast. It would take most of the nineteenth century for the West to be won, but the foundations of modern North America were laid in the mismanaged French sugar colonies of the Caribbean, and in the French willingness to pass up on Liberty, Equality and Fraternity, when those ideals stood between them and making sugar profits.

All the same, the French did not lose out entirely. In November 1814 the island of Guadeloupe, which Napoleon's France had also lost to England, was returned to newly royalist France. On duty there, Edward Codrington (later Admiral Sir Edward Codrington) noted sardonically in his *Memoir* that:

> The people of this island will, I suspect, have cause to regret the change in their government, for there are already sixty clerks come to do what eleven have done under us . . . and the whole of these . . . are to be paid by the islanders who will be taxed enormously, and be unable to profit by the permission to purchase negroes.

Some of the people of the island had even more immediate regrets, for unlike the people of Haiti, those on Guadeloupe who had been slaves were to become slaves again, as Codrington noted in another letter:

> . . . the spot was pointed out to me as the last retreat of those who were struggling for liberty. A considerable number of people of colour who saw no hope left but a return to slavery, by joint consent blew themselves up together rather than ask life upon such degrading terms.

In the peace after Waterloo, the British sugar growers on Jamaica and the smaller islands found themselves faced with plummetting sugar prices. They also knew the slaves had a role model to look to on Haiti, a promise of freedom that might one day be seized, for even if the hated slave trade had been stopped, slavers were still sailing the oceans under the flags of other nations, and the slaves were still slaves. England needed the profits, and everybody needed their sugar, just as Ralph Clark did.

SUGAR, COFFY AND RALPH CLARK

Ralph Clark comes into our view first as a marine who sailed for Sydney on the convict transport *Friendship* when Arthur Phillip's First Fleet set out to create a settlement at Botany Bay. He, and the rest of the fleet, faced uncertain prospects on the other side of the world from 'home', and he quickly realised that some things would have to change. He wrote in his journal, in his idiosyncratic spelling:

> . . . to day for the first time in my life drinked my tea without Sugar which I intend to doe all the Voyage as my Sugar begins to grou Short therfor will only drink tea and Sugar now and after we get on Shore on certaind days . . .

Clearly, the fleet carried some sugar, but this seemed to be mainly as a form of medicine, because it is mainly the surgeons who seem to have mentioned it in surviving records. Surgeon John White complained that his first hospital in Sydney had no sugar, sago, barley, rice or oatmeal, all essential hospital supplies, but Arthur Bowes Smyth, surgeon on the *Lady Penrhyn*, records how, on the journey out, he gave 'a quantity of Sagoe & soft sugar to every Birth [berth] of the Women as an indulgence, as I had plenty of both by me'. In fact, it was not until *Lady Penrhyn* was almost to China, on the journey back to Europe, that he recorded running out of sugar, just days before dropping anchor at Macao, where he was able to replenish his stocks.

Clark, though, was less fortunate. The commander of the First Fleet, and Governor, Arthur Phillip, had collected some sugar cane at the Cape of Good Hope, and while some seems to have been tried unsuccessfully in Sydney, most of it was sent on to Norfolk Island, to the north-east of Sydney. Supposedly, the sugar cane did well, or so Lieutenant King said. In a surviving letter, Clark notes darkly that Lieutenant King had written to Sir Joseph Banks about how he had made both sugar and rum, but the marines saw none of it, and Mr King, said Clark, 'would not make his dispatches Public, because he knew they did not agree with the Private Accounts'.

There is a whiff of scandal here, but there was to be no whiff of sugar, molasses or rum for the poor marine lieutenant. In early 1791, Clark wrote in his journal that he yearned for tea and sugar, adding that he had not had tea or wine for six months. There was no coffee, either:

> . . . our Breakfast is dry bread and Coffy made from burnt wheat and we are glad even to be able to get that—God help use I hope we will Soon See better days Soon for the[y] cannot well be Worse.

A week later, with hope that a ship was about to arrive, he repeated his hope for tea and sugar, but it seems he had to wait until May for any sugar. As early as 1788, Clark had written to a brother officer at Plymouth, asking for

> Viz: 6 or 8 lb of Tea, about 40 or 50 lb of Sugar, 6 lb of Pepper, 2 pices of printed Cotton at about 3 or 4 dollars a pice for window Curtains and a dozen the Same Kind of plates as You gave me and let me know what the cost and I will Send you ane order for the Same—be So good as to make my best and tendrest wishes to Mrs. Clark and inform her I have wrote her by this opportunity . . .

While Clark was roughing it in the sugar-free wilds of Australia and Norfolk Island, where he had a daughter by a convict girl named Mary Branham, his wife Betsy was at home in England, caring for his young son, also called Ralph.

Clark returned to England in 1792, and left a pregnant Betsy in England when he sailed off to fight in Haiti in May 1793, taking young Ralph with him to the sugar islands of the Caribbean. There, Clark senior died in a naval action against the French, while his son died, apparently on the same day as his father, of yellow fever; a little earlier, Betsy had died after giving birth to a stillborn child.

Perhaps he died happy though, because in June 1794 he wrote a letter describing how they had taken '45 Ships, 36 of them are large Ships, deeply loaded with Sugar, Coffy, Cotten and Indigo'. We can only hope that he got his fill of the captured Coffy and Sugar before he was cut down, leaving behind his incomplete papers, his name on a small island in Sydney Harbour where he once had a garden, and possibly his daughter by Mary Branham, of whom we know nothing more.

SUGAR BECOMES A COMMONPLACE

By the time Ralph Clark died, most Europeans knew and hungered after sugar. The percolation of sugar down the social scale had been slow but steady. In 1513 the King of Portugal sent sugar effigies of the Pope and twelve cardinals to Rome as a mark of his esteem, and in 1515 sugar was taken to King Ferdinand on his deathbed. In 1539 Platine recorded the French proverb: *Jamais sucre ne gâta viande* ('adding sugar never hurt any food')— which only referred to the food of the rich, but that was changing fast. At the end of the sixteenth century, Queen Elizabeth and her courtiers may have rotted their teeth, but sugar had already extended beyond the Court circle. In England in the mid-1800s, when Dickens was writing his novels, sugar in one's tea was a commonplace throughout British society.

The price of sugar probably tells the story of its filtering down through the classes better than anything else. In modern terms, a kilogram of sugar cost about US$24 in 1350–1400, $16 in 1400–50, $12 in 1450–1500, and $6 in 1500–50, when the first Brazilian sugar reached Europe, and now it was cheap enough for ordinary people to aspire to enjoy it. Warfare, cane disease and weather might cause small fluctuations, but the price continued to fall. The fall in price was more than matched by the increased demand, and that demand was matched by a continued growth in the area under cultivation with sugar cane.

The sack of Antwerp in what is now Belgium, by Spanish forces opposing Dutch independence, led to the destruction of the refineries there and by 1600 the English, French and Dutch were all in the sugar refining business. The Antwerp workers dispersed, taking their knowledge with them to London, Amsterdam, Hamburg and Rouen, resulting in increased competition for raw sugar to feed the new refineries. All the same, the English were slow to begin growing sugar in their colonies. As Richard Ligon tells us, most of them started as growers of tobacco, cotton, indigo and ginger, along with cassava, plantains, beans and corn. This was the usual island pattern, with sugar only coming in later.

Barbados, for example, only started planting sugar in about 1640, and had a number of poor years, until the arrival of Dutch and Jewish refugees ejected from Pernambuco in northern Brazil. They provided the technical knowledge to make more and better sugar. The disruption in Brazil had also reduced the amount of sugar available in Europe, making this an excellent time for Barbados to adopt the new crop.

A simultaneous plus and minus was that Barbados had many smallholdings already cleared. This made it easy to plant cane, but as the best results financially came with a mill for every 40 hectares (100 acres) of cane, small farmers could not get capital, and their farms were quickly swallowed up by their larger neighbours. Many landless people joined the 1655 British invasion of Jamaica, and gained land there, but the surviving Barbadian planters were now rich and powerful, thanks to sugar.

Between 1663 and 1775, English consumption of sugar increased twentyfold, and almost all of it came from the Americas. Sugar was big business, and led to the first *Molasses Act* being passed in Parliament in 1733. It was set to last for five years, but it was regularly renewed, consolidated in the *Sugar Act* of 1764, and only repealed in 1792. This Act, in its various forms, set a duty of 5s. per hundredweight on sugar, 9d. per gallon on rum and 6d. per gallon on molasses brought into a British colony from a foreign source, while the importation of French produce into Ireland was forbidden. The result of the tax was to encourage wholesale smuggling of sugar and molasses into the thirteen British colonies in North America (and their eventual revolt). It seems not to have been worth the effort to smuggle sugar into England, however. In 1852, Captain Landman, late of the Royal Engineers, recalled an incident at Plymouth in 1796:

> [We walked] towards the harbour, and on our way met an immense number of thin women proceeding with the utmost expedition, whilst all those we overtook, about equal in number, were large stout females, evidently waddling along with difficulty. On seeing these, Phillip explained that the latter were all wadded with bladders filled with Hollands gin, which they manage to smuggle under these dresses, whilst the others were thin and light, having delivered their cargoes at the waterside . . . everybody knew the trade they were engaged in.

Perhaps sugar was just too hard to carry, but more probably it was too bulky to repay the time, risks and effort, so people in Europe consumed taxed sugar and put up with it. Even with taxes, everybody had to have their sugar.

By 1675, England was seeing 400 vessels, each carrying an average of 150 tons of sugar, arriving from the colonies each year. France was exporting equally large amounts of sugar from its colonies. All sorts of arguments were proposed against the tax on sugar, from the suffering of the planters and their slaves to the welfare of Britain, but the French and British governments kept on taxing.

The hunger for sugar was by no means an English phenomenon. When Tobias Smollett travelled in France and Italy in 1766, he described tea at Boulogne: 'It is sweetened all together with coarse sugar, and drank with an equal quantity of boiled milk.' He recorded better Marseilles sugar at Nice, but complained that the liqueurs there were so sweetened with sugar as to have lost all other taste. He complained also of the flies, which apart from anything else, 'croud into your milk, tea, chocolate, soup, wine, and water: they soil your sugar'. Later, Smollett referred to buying sugar and coffee in Marseilles, giving us the trinity of the sugar promoters: tea, coffee and chocolate, all needing to be sweetened.

Doctor Johnson also had something to say on the subject of the French way with sugar according to Boswell:

> At Madame ——'s, a literary lady of rank, the footman took the sugar
> in his fingers, and threw it into my coffee. I was going to put it aside;
> but hearing it was made on purpose for me, I e'en tasted Tom's fingers.

Although it has been recorded as replacing honey in recipes for chardequynce, a sort of jam or compote, as early as 1440, sugar only began to appear regularly in recipes about the mid-1700s. Mrs Hannah Glasse published her *The Art of Cookery Made Plain and Easy* in 1747, and the book revealed sugar as a standard item in the kitchen—one cake recipe calls for 'three quarters of a pound of the best moist sugar'.

When Thackeray wrote of the England of the early nineteenth century in *Vanity Fair*, he used sugar grades to distinguish social classes. Here he describes Mr Chopper:

> The clerk slept a great deal sounder than his principal that night; and, cuddling his children after breakfast (of which he partook with a very hearty appetite, though his modest cup of life was only sweetened with brown sugar), he set off in his best Sunday suit and frilled shirt for business, promising his admiring wife not to punish Captain D.'s port too severely that evening.

The purer white sugar that would satisfy others was out of the reach of a clerk, but tea and sugar had become essentials of life. Charles Dickens mentioned sugar 102 times, on thirteen of those occasions in the phrase 'tea and sugar', while rum rated almost 150 mentions. Dickens makes it clear in a scene in *Nicholas Nickleby* that it was not uncommon for a cook to expect tea and sugar to be provided in the terms of employment.

The habit of drinking tea was a key factor in saving lives when cholera reached Britain in the nineteenth century, because boiling water to make tea killed the germs in the water. The infamous Broad Street pump in London's Soho, drawing water from a well lying beside a cesspit, killed most of the non-tea drinkers in the area before Dr John Snow made the link between the pump and cholera, and called for the removal of the handle on the pump.

Wherever the English went, whether aristocrat or commoner, tea and sugar went with them. In 1826, convicts in Australia had a weekly ration of 7 lb. meat, 14 lb. wheat and 1 lb. sugar; in 1830, shepherds and other farm employees were allowed

10 lb. meat, 10 lb. flour, 2 lb. sugar, and 4 oz. tea each week. In 1833 a teenager called Edward John Eyre set off to make his fortune in the Australian bush, with a quart pot, a little tea and sugar, and some salt.

On his first night, Eyre and his dray driver shared a glass of rum before settling down to sleep for the night under the stars. Within a decade Eyre would be winning fame as an explorer, a finder of paths across the trackless Australian landscape; within three decades he would be reviled by much of England and most of Jamaica as the man who hanged more than 400 former sugar slaves on Jamaica. In 1833 the teenage boy was an ordinary person, but sugar and rum were staples of life for an Englishman in the colonies.

SUGAR

Hazardous Properties: It is well known that sugar refiners face an industrial hazard which consists of an irritation to the skin; it may be a form of dermatitis. Bakers also experience a dermatitis due to sugar.

Storage and Handling: Personnel who must work with this material continually and who are sensitive to it should wear protective clothing to avoid skin contact.

N. Irving Sax, *Handbook of Dangerous Materials*, New York, 1951

6
A SCIENCE OF SUGAR

Henri Louis Duhamel du Monceau, the French agricultural encyclopaedist, recorded that the 'Jamaica Train' was probably invented about 1700, but like all other advances in the art of making sugar, it was slow to spread. The Portuguese and the Spanish knew about simple sugar making because they had long been involved in making it on the Atlantic islands; it had given them a head start in the sugar industry in the Americas, but they seem to have given little thought to advancing their skills.

English planters were using the Jamaica Train around 1700. By 1725, the French were using what they called the English Train, the Cubans adopted the French Train in 1780, and the Brazilians brought it into use about 1800. The early forms of the Jamaica Train had four large pans, each 4 feet (120 cm) in diameter with a flat bottom, set into masonry so the flames hit only the base. This meant that as long as there was a small amount of liquid in the pan there would be no charring of sugar up the sides, as happened with the old conical pans. Later forms had five pans of decreasing size, but in each case a single fire heated all the pans, with the draught being carried along beneath them.

The pans were all heated with stone coal, not charcoal, du Monceau explained, before diving into the complexities. The sugar was treated with lime water, then a clearing medium of egg white or bullock's blood was added (the English called this 'spice'). The ratio was about 80 eggs or 2 gallons of blood to 4 tons of sugar. He noted that isinglass (a form of gelatin obtained from fish) did not work as well as blood to clarify the solution. When the scum rose to the surface, he wrote, the fire was drawn (reduced or damped down), and after fifteen minutes the scum was scraped off. This would be repeated, until a clear bright liquor was obtained and strained through a blanket.

The materials used to clarify sugar at various times have included wood ash, milk, egg white, blood, charcoal, lime, sulfurous acid, phosphoric acid, carbon dioxide, alum and as we have seen, lead acetate. Marco Polo referred to the use of wood ash, lime and alum as an Egyptian practice.

Du Monceau tells us that the juice was put into a boiling vessel called either a clarifier or a racking copper, usually holding 500 gallons (more than 2000 litres), where the temperature was raised to about 175 degrees on de Réaumur's scale (in our terms, 285°F or 140°C). At this point, milk of lime was added to coagulate impurities, which would rise to the top as a scum. When the scum 'cracked', a cock in the bottom was opened to drain the cleared liquid, leaving the 'mud' behind to be used in the making of rum.

Getting the amount of lime right was always a challenge to the sugar masters. If the liquid was not tempered properly, the sugar would not crystallise, but if too much lime was added, the liquor would turn green as chlorophyll entered the solution, and would later form a dark sugar with a great deal of molasses and an unpleasant smell. The answer was to turn to science, and Dr John Shier in British Guiana introduced the use of Robert

Boyle's litmus test. When the litmus just turned from red to blue, enough lime had been added.

Later, Francis Watts on Antigua replaced litmus with phenolphthalein (hopefully in small amounts or later removed, as phenolphthalein is a laxative), and by 1870 phosphoric acid was being used to remove the excess lime, producing a calcium phosphate precipitate. Reflecting a higher than usual awareness of the needs of the soil, an early Australian refinery began soon after to convert this precipitate to superphosphate and return it to Madagascar in the ships that had brought the raw sugar across the Indian Ocean.

In the 1760s butter was added to the boiling pan (in place of Ligon's Sallet Oyle), and the mixture was boiled to the strike point. This strike point is a matter of judgment and testing: a small amount of the liquid is drawn out between finger and thumb, revealing its state to the sugar master. If the thread breaks near the top, the liquid needs more heat, but if it breaks near the bottom it is ready.

When it had reached the strike point, the liquid went first to a cooling vessel, and then was poured into a number of conical earthenware pots with holes in their pointy tops and open at the base. These were allowed to stand upside down for about six days while the mother liquor ran off, after which the sugar was usually 'clayed' until it was a satisfactory colour. The loaves of sugar were then stood in a hot room to dry. Claying was preferred, because this added value to the crop in the colony, and gave a better return to the plantation owner.

There is a pretty legend that claying was discovered when a chicken wandered through a curing house and stepped across some muscovado sugar, leaving a trail of white footprints behind, thus inspiring people to clay sugar to make it white. This tale is widely told but improbable, because it has little to do with the claying process. When the cone full of muscovado was turned over and allowed to drain, it reached a point where most of the viscous part had drained from the pot, and the rest clung tenaciously to the crystals. Claying involves tapping the upside-down cone gently to settle the sugar, covering the open base with clay and then adding water so that it drips slowly down and out the hole, leaching out the remaining molasses so that it can be collected and sent to the rum makers, or boiled again to obtain more sugar.

When the cone was turned back up and bumped gently, the result was a sugar loaf sitting on a clay base: a cone of sugar whitest at the bottom, and brownest at the top. The problem was that the home governments, always seeing the colonies solely as a source of delivering raw materials and potential profit to the home country, placed taxes on clayed sugar, which effectively reduced the profit to the plantation.

Detailed as they were, du Monceau's descriptions of the intricacies of sugar making, so urgently needed by the sugar growers, were nowhere near as memorable as the educational efforts to be found in the strange poetic outpourings of Dr James Grainger.

THE GEORGIC DOCTOR

James Grainger, poet, author, physician and educator of sugar planters, began his career as a surgeon in the 13th Foot during the Jacobite rising of 1745. He later set up practice as a doctor in London, and was a contributor to different literary journals.

Perhaps to develop an image as a Renaissance man, he published a study of army diseases, but John Pringle's study was superior, and came out the year before Grainger's. That, and the fact that Grainger wrote in Latin, while Pringle had published in English, meant Grainger's book lost by comparison. Still, he mingled with and was admired by the best literary people in London—the likes of Samuel Johnson, James Boswell and Thomas Percy—and he had a famous battle with Smollett over his translation of the poems of Tibullus, among other things.

In 1759 Grainger set sail with his patron, John Bourryau, to visit Bourryau's sugar plantation on the island of St Kitts. There was a war going on with France at the time, so the ship sailed in convoy, and on the way out Grainger was transferred to another ship to treat a woman stricken with smallpox. So it was that he met the lady's daughter, who went by the improbable name of Daniel Mathew Burt. Her name indicated nothing of her gender, but did serve to remind other Kittitians that she was connected to both the Daniel and Mathew families, both prominent on the island of St Kitts.

Shortly afterwards, James Grainger, MD and literary lion, married to the former Daniel Burt, was practising medicine on St Kitts and looking for some activity suited to his talents, whereby to make his reputation and his fortune. It was clear that the planters needed instruction in the arts of sugar making, so this became his project. Rather than straightforward prose, however, he found a model in the *Georgics* of Virgil, which describe agricultural practices in Rome in the first century BC, including the keeping of bees.

This style, common enough at the time, was shortly to go out of fashion (though Erasmus Darwin, Charles' grandfather, became famous when he used it to write about nature, classification and his views about the evolution of new species). Erasmus was highly rated as a poet, and influenced many of the poets of early nineteenth-century England, but he made it foolishly obvious that when writing Georgics, one does not call a spade a spade. Instead, it must be called the:

> Metallic Blade, wedded to ligneous Rod
> Wherewith the rustic Swain upturns the sod.

After his squabble with Smollett, Grainger should have been more wary of giving people an easy chance to make fun of his style, but he appears to have learned only slowly. *Sugar-Cane: A poem in four books*, published in London in 1764, was a massive blank verse presentation on all aspects of cultivating the sugar cane, caring for slaves, making sugar and much more. Sad to say, the planters did not care for the elegant language, while many Londoners could not see the need to address in careful detail such important issues as the rats that destroyed the cane, the diseases that affected slaves, or manure. People like Dr Johnson got many a belly laugh at Grainger's expense, according to Boswell:

> He spoke slightingly of Dyer's *Fleece*.—'The subject, Sir, cannot be made poetical. How can a man write poetically of serges and druggets? Yet you will hear many people talk to you gravely of that excellent poem, *The Fleece*.' Having talked of Grainger's *Sugar-Cane*, I mentioned to him Mr. Langton's having told me, that this poem, when read in manuscript at Sir Joshua Reynolds's, had made all the assembled wits burst into a laugh, when, after much blank-verse pomp, the poet began a new paragraph thus:—
>
> 'Now, Muse, let's sing of rats.'
>
> And what increased the ridicule was, that one of the company, who slily overlooked the reader, perceived that the word had been originally MICE, and had been altered to RATS, as more dignified.

That was almost certainly untrue. As Richard Ligon had made clear, rats have always been a problem in the cane fields, both because they damage the canes and because they spread a nasty disease called leptospirosis. Still, by twisting things a little, Johnson also had his fun with the work, as Boswell tells it:

> Johnson said, that Dr. Grainger was an agreeable man; a man who would do any good that was in his power. His translation of Tibullus, he thought, was very well done; but *The Sugar-Cane, a poem*, did not please him; for, he exclaimed, 'What could he make of a sugar-cane? One might as well write the "Parsley-bed, a Poem"; or "The Cabbage-garden, a Poem".'

Grainger may have hoped that this work would make his literary reputation and earn him enough money to retire 'home' to enjoy his riches. On St Kitts he doctored slaves, attended to the plantocracy—the 'Creoles', as they were known locally—and sold medicines as well. He also travelled to nearby islands when called upon to treat the sick. His poem had a serious message, however—it was intended to enrich the new planter by teaching him all the necessary arts such as:

> What soil the Cane affects; what care demands;
> Beneath what signs to plant: what ills await;
> How the hot nectar best to christallize;
> And Afric's sable progeny to treat:
> A Muse, that long hath wander'd in the groves
> Of myrtle-indolence, attempts to sing.

Most amusement seemed to be aroused by his paean of praise to the marvels of good compost (which still finds favour with gardeners today):

> Of composts shall the Muse descend to sing,
> Nor soil her heavenly plumes? The sacred Muse
> Naught sordid deems, but what is base; nought fair
> Unless true Virtue stamp it with her seal.

> Then, Planter, wouldst thou double thine estate;
> Never, ah never, be asham'd to tread
> Thy dung-heaps, where the refuse of thy mills,
> With all the ashes, all thy coppers yield,
> With weeds, mould, dung, and stale, a compost form,
> Of force to fertilize the poorest soil.

While compost and manure might have been important to the farmer and his plantation readership, they were unlikely to appeal to his potential London audience, and the sales of his work were poor. But there was more:

> Whether the fattening compost, in each hole,
> 'Tis best to throw; or, on the surface spread;
> Is undetermin'd: Trials must decide
> Unless kind rains and fostering dews descend,
> To melt the compost's fertilising salts;
> A stinted plant, deceitful of thy hopes,
> Will from those beds slow spring where hot dung lies:
> But, if 'tis scatter'd generously o'er all,
> The Cane will better bear the solar blaze;
> Less rain demand; and, by repeated crops,
> Thy land improv'd, its gratitude will show.

If compost was important, the slaves could not have been too enthusiastic about it. To prevent sheet erosion on the cleared ground, cane was planted in 'holes', squares about 1.5 metres across and 15 cm deep, which made it difficult or impossible to get carts in when the manure was needed, as the young cane shoots began to appear. The manure had to be carried in on the slaves' heads in baskets, as Miss Schaw described it in her journal in 1774, a few years after Grainger died:

> Every ten negroes have a driver, who walks behind them, carrying in
> his hand a short whip and a long one ... When they are regularly

ranged, each has a little basket, which he carries up the hill filled with
the manure and returns with a load of canes to the Mill. They go up at
a trot and return at a gallop . . .

The 'little basket' with its contents probably weighed 35 kilo-
grams (75 pounds or more), and its soggy contents would have
been dripping down on the carrier the whole time. It was
certainly the task most resented by the slaves who were inter-
viewed after emancipation, so a nervous manager might be
tempted, for fear of poison in his food, to forget about the
manuring of the fields.

It did not help the cause of scientific farming when, around
1816, Lord Dundonald wrote a 'treatise on Chymistry as applied
to agriculture' which recommended peat as the best manure for
cane. As Thomas Spalding commented, 'He does nothing but
what a mind heated to excess would have thought of, when he
recommends that peat should be prepared in Scotland and sent
to Jamaica for the purpose.'

Like Dundonald, Grainger failed because he failed to stick to
the simple facts. Many of the islands had poor soil, and without
compost the cane eventually grew poorly because vital elements
were not returned to the soil. In some parts of Brazil the planters
could move to new land as the old land was exhausted, but on
the smaller islands this was simply not possible.

Whatever the reason, the sophisticated people of London
laughed at Grainger's attempts to instruct, and he never reaped
the hoped-for reward from his efforts. English readers missed
what he meant in his oblique descriptions, while those in the
islands who might have benefited either thought they knew it
all, or were far from bookish.

The medical side of the poet comes to the fore when Grainger
writes of caring for slaves. As far back as 1717, Lady Mary

Wortley Montagu had written from Constantinople to describe how the Turks dealt with smallpox by infecting themselves when they were healthy:

> The small-pox, so fatal and so general among us, is here entirely harmless, by the invention of *ingrafting*, which is the term they give it. There is a set of old women who make it their business to perform the operation every autumn, in the month of September, when the great heat is abated. People send to one another to know if any of their family has a mind to have the small-pox . . .

The ingrafting, or variolation, was carried out with smallpox material from survivors, and so would have been a slightly weaker strain of the virus. This contributed to its successful use in Boston by Cotton Mather before his death in 1728—and he claimed to have learned about the practice from African slaves. Even George Washington arranged to have his troops variolated before they went into battle.

Later, it was this practice of variolation which allowed Edward Jenner to test the effect of cowpox, when he first gave the young James Phipps the harmless cowpox and then followed the completely routine practice of variolation. The common modern claim that Jenner was in some way unethical in 'deliberately giving a young boy smallpox' is based on gross ignorance of the medical norms of his day.

But if we have forgotten about ingrafting now, it was well known back in Grainger's time, which makes it odd that he should feel the need to advise his readers of what they should have known:

> Say, as this malady but once infests
> The sons of Guinea, might not skill ingraft
> (Thus the small-pox are happily convey'd;)
> This ailment early to thy Negroe-train?

Grainger also offered sound advice on industrial safety that might have helped Macandal a few years earlier:

> And now thy mills dance eager in the gale;
> Feed well their eagerness: but O beware;
> Nor trust, between the steel-cas'd cylinders,
> The hand incautious: off the member snapt
> Thou'lt ever rue, sad spectacle of woe!

He also provided advice on allowing the slaves to drink cane juice during the harvest:

> While flows the juice mellifluent from the Cane,
> Grudge not, my friend, to let thy slaves, each morn,
> But chief the sick and young, at setting day,
> Themselves regale with oft-repeated draughts
> Of tepid Nectar; so shall health and strength
> Confirm thy Negroes, and make labour light.

And he inveighed against wicked Frenchmen who adulterated their sugar, something no true Britisher would do (though G. K. Chesterton seemed to think English grocers did it all the time). According to Grainger:

> False Gallia's sons, that hoe the ocean-isles,
> Mix with their Sugar, loads of worthless sand,
> Fraudful, their weight of sugar to increase.
> Far be such guile from Britain's honest swains.

If the planters had paid more attention to Grainger's sound advice, or if Grainger had couched his sound advice in less complex terms, how many lives might have been saved? Grainger wrote on a variety of worms and their treatment, depression, nutrition and more, all directed at keeping slaves healthy and working—but above all, alive. Grainger even provided notes to the work, to explain, for example, that '[T]he mineral product of the Cornish mine' was in fact tin, which he pointed out could be used as a vermifuge (a treatment for worms) in either powder or filings form.

In the end, Grainger's poem fell badly between two stools. By 1860 George Gilfillan could include Grainger among the 'less-known poets'. In 1930 he was included in *The Stuffed Owl*, an anthology of bad verse. Neither of these was quite as damning as it sounds: Gilfillan's list of lesser known poets included John Donne, Sir Philip Sidney, Christopher Smart and Jonathan Swift, and the alleged bad versifiers in *The Stuffed Owl* included Burns, Byron, Keats, Longfellow, Smart again, and Wordsworth.

All in all, not bad company in which to find a Scottish doctor on the make.

FOR WORMS

Give a child one year old 15 drops of spirits of turpentine on sugar, fasting, for three mornings in succession; follow the last dose with a good dose of castor oil; this forms an excellent vermifuge. The dose of spirits of turpentine for a child two years old is 20 drops, three years old 25 drops, four years old 30 drops, &c.

Daniel Young, *Young's Demonstrative Translation of Scientific Secrets*, Toronto, 1861

7

RUM AND
POLITICS

Approval given to Mr Waterhouse to supply King James's ships at
Jamaica with Rumm instead of Brandy, he takeing care that the good
or ill effects of this proof, with respect as well to the good Husbandry
thereof as to the Health and Satisfaction of our Seamen, be carefully
inquired into by you and reported to us within a yeare or two (or
sooner if you find it necessary for our further satisfaction in the same).

Samuel Pepys, Secretary to the Navy, 3 March 1688

Throughout the ages, wherever people have looked to add some
intoxicating interest to life, they have remembered or rediscovered the art of getting zing from sugar. When Captain Cook
took his men to Hawaii in 1779, he described his way of treating
cane juice:

Having procured a quantity of sugar-cane, and finding a strong decoc-
tion of it produced a very palatable beer, I ordered some more to be
brewed for our general use. But when the cask was now broached, not
one of my crew would even so much as taste it. I myself and the officers
continued to make use of it whenever we could get materials for
brewing it. A few hops, of which we had some on board, improved it

much. It has the taste of new malt beer; and I believe no one will doubt
of its being very wholesome. Yet my inconsiderate crew alleged that it
was injurious to their health.

The ship had taken on four casks of rum at Rio, so maybe the
men wanted none of the 'beer' because there was still rum left.
Or maybe they thought it was just another of Cook's cures for
scurvy, like his 'portable soup' and 'sour krout'. Christmas was
approaching, and that was a time when Cook's crew would nor-
mally make merry, as they did on his first voyage to the South
Sea. Here is part of Joseph Banks's journal entry of 25 December
1768, just ten years earlier, which suggests that on that voyage
at least, Cook's ship had enough liquor for all:

Christmas day; all good Christians that is to say all hands get abom-
inably drunk so that at night there was scarce a sober man in the ship,
wind thank god very moderate or the lord knows what would have
become of us.

Cook's official account was more understated: 'Yesterday being
Christmas day the People were none of the soberest.' Neither
Banks nor Cook indicates what the liquor was; rum was used on
ships earlier, but it was only in 1775 that it became the standard
issue liquor for sailors in Britain's navy. Before then, a variety of
alcoholic drinks were in common use, so we find Banks record-
ing that when the crew of *Endeavour* 'crossed the line' (passed
over the equator for the first time), the first-timers could accept
being ducked, or 'give up 4 days allowance of wine which was
the price fixd upon'. But standard or not, rum was a common
tipple for British sailors for a long time.

Distillation is the oldest chemical craft in the world—the
earliest surviving piece of chemical equipment, a distillation
apparatus for separating perfume ingredients, has been dated
to 3600 BC. Islamic chemists knew all about distillation (we

get our word 'alcohol' from Arabic), and had taken the knowledge to Spain. So while rum is usually associated with the Caribbean or the Americas, it is quite possible that an alert Spaniard had noted that sugar juice, left to stand, was in the habit of fermenting into Cook's 'beer'; from there it would be but a small step to producing some rum-like liquor on the quiet.

Although the origin of its name is obscure, rum has been known since the English settled in Barbados in 1627, and the Spanish and Portuguese were possibly involved in distilling spirits on their sugar plantations even earlier than this. The art of distillation is often said to have come into the islands with the Jewish refugees from Brazil, who had learned to make *cachaça*, which is distilled from the raw cane juice rather than from molasses, as rum is. The only snag here is that the dates do not add up, since the Jews did not flee Brazil until around 1654, by which time rum was common on Barbados.

The *Etymological Dictionary of the English Language*, published in 1888, suggests rum is a corruption of *brom*, a Malay word said to mean arak, the spirit distilled from palm juice, but this is not so. That word is *beram*, and while this would be pronounced rather like b'rum, the Malays' *beram* is actually brewed from rice or tapioca. This is a long way from rum which, according to the Jamaican Excise Duty Law No. 73 of 1941, can be described as 'spirits distilled solely from sugar cane juice, sugar cane molasses, or the refuse of the sugar cane, at a strength not exceeding 150 per cent proof spirit'—which means 75 per cent alcohol.

To Richard Ligon, rum was a 'a hot, hellish and terrible liquor' also known as kill-devil, and before long the French called it *guildive*, corrupted without understanding from kill-devil, while the Danes called it *kiel-dyvel*.

Rum today is produced from sugar cane by yeast fermentation. The 'wash' that is produced is about 6 per cent alcohol, and this distils to a clear, colourless liquid with up to 80 per cent alcohol and a sharp taste. Commercial white rums are essentially this product diluted back to 40 per cent alcohol, while gold rum is the same product after it has been aged in small (40 gallon) oak barrels. The ageing process sees some of the pungent volatile components evaporate, while chemical reactions between the rum and the oak add flavour. As well, some oxygen probably finds its way in to convert some of the alcohol to aromatic esters, compounds which give a variety of 'fruity' tastes.

If sugar made the planters a profit, it was rum that sustained them. Even when sugar prices fell, rum was there to provide cash income, and if the price of rum fell it could be stored, or used to drown the planters' sorrows. It could also serve to sweeten a sailor's harsh life, a life that Dr Johnson likened to being in gaol, enlivened with the prospect of being drowned.

The chorus 'fifteen men on the dead man's chest, yo-ho-ho and a bottle of rum' reminds us that rum was one of the few creature comforts available to pirates, especially when they were at sea. Captain Kidd is said to have landed fifteen crew on the smallest of the Virgin Islands (supposedly coffin-shaped) to bury his treasure, and killed them all to keep his secret safe. Perhaps he gave them extra rum, or poisoned it to make them a pushover, and then pushed them into the hole—there were many ways rum could kill a sailor.

'OLD GROG'

Admiral Edward Vernon, born in 1684, was a captain in the Royal Navy at 21, and a rear-admiral at 24. He served successfully for many years until (according to his supporters) he was forced out because he was right too often. By then he had won many famous battles, but none so famous as his battle against rum for sailors. Too many Jack Tars went aloft, he argued, with too much good Jamaica rum under their belts, made a false step—and died.

'Good Jamaica rum', Vernon may have called it, but all too often it was the distillers' bilge water, lees and rubbish that merchants could not dispose of in any other way, sold on to conniving clerks, landsmen leeches who were ostensibly in the pay of the Admiralty but simultaneously were more lucratively in the pay of the merchants. These men grew rich on the payments they had for taking third- and fourth-rate stuff.

At least the sailors could test the rum for the amount of spirit. They just poured some on a pinch of gunpowder and set a spark to it—if the rum was proof or better, the water in it boiled off as steam, the powder dried, there was a flash as the powder burned, and the rum passed the proof, or test. Underproof rum left the powder damp and unfired.

Thus the sailors could make sure the rum contained enough spirit to warm a man's belly—but that was also enough to make him careless-footed when aloft, and slow to react in battle. Before the 1650s, wine and beer were given to sailors, then brandy was used for a while, but by the 1680s, after Britain took Jamaica from the Spanish and expanded sugar production, somebody had to use all the rum that was being made. Since rum, unlike wine and beer, did not go bad at sea, it was added to the range of acceptable beverages.

Still the rum killed men, so Admiral Vernon ordered that it be watered before it was issued. The men already called him 'Old Grog', on account of the waterproof boat cloak he wore made of a cloth called grogram or grosgrain; by extension, the watered rum gained the name of grog. The word passed into the English language, even as the sailors complained that Old Grog was depriving them of an essential of life.

In 1740 Vernon ordered that the daily allowance of one pint of rum per man be mixed with one quart of water in a scuttled butt, a barrel with one end removed kept for that purpose. This was to be done on deck, in the presence of the lieutenant of the watch, who was to see that no man was cheated of his proper allowance.

Grog could also be used as a reward for sailors who carried out complex and difficult tasks. The mainbrace was a fearsomely heavy cable which controlled the mainsail, and if this parted and needed to be joined, a long splice was required, a form of joining that would allow the cable to pass through the blocks (pulleys to landsmen). Until the mainbrace was spliced, the ship had to be held on one tack, so the men who carried out this task needed to work fast and well, and thus earned an extra ration of grog. When there was a general bonus issue, the crew was also said to 'splice the mainbrace'.

Where rum was used on ships after Vernon wrote his order, throughout the navy the daily allowance of one pint of rum mixed with one quart of water was issued in two parts: one in the forenoon and the other in the evening. In 1824, the evening issue was stopped and in 1850 the ration was cut to one gill (one-eighth of a pint) of rum with two gills of water per man per day. In 1937 the amount of water was halved, and in 1970 the 'grog era' ended when the Royal Navy's rum ration was cut altogether.

RUM, SUGAR AND TAXES

In the eighteenth century, rum was important wherever English was spoken. While Longfellow's story of Paul Revere's ride may have glossed over his stop at Medford, the famous silversmith and equestrian was more open about it, and recalled later that he 'refreshed himself' at Medford, leaving no doubt that he had taken some of the New England rum centre's most famous product. George Washington is reputed to have gained election to the Virginia House of Burgesses as a result of his judicious sharing of 75 gallons of rum among the voters.

In the infant colony of Australia rum was important because the colony at Botany Bay was run by naval men while the soldiers who guarded the convicts were marines. Both naval officers and marines (a separate service) had a high regard for rum, and the marines especially saw it as a way to make money. Rum became a major item of currency in the settlement at Sydney Cove. William Bligh, of *Bounty* fame, and later the governor of the colony, noted:

> A sawyer will cut one hundred timber for a bottle of spirits—value two shillings and sixpence—which he drinks in a few hours; when for the same labour he would charge two bushels of wheat, which would furnish Bread for him for two months.

Convinced that rum was an evil, Bligh ran foul of the marines—known locally as the Rum Corps—who had gained a stranglehold on the small colony. He seized an illegally imported still and provoked the Rum Rebellion, a mutiny which should by rights have led to the hanging of those responsible, but did not. The Rum Rebellion, and the English government's varied responses to it over two decades, led in the end to the colony of New South Wales gaining a measure of self-rule.

Rum and sugar meant money, and money meant power, so rum and sugar influenced the way power was used, and not just by way of dispensing strong spirits to the voters. Because of the wealth it brought, and the way that wealth was used, the sugar trade can be credited with (or blamed for) the large number of people of African origin in the Americas and Britain today, the spread of English-speaking influence across North America, and much more. It is even possible to make the case that today's political world was shaped by events and forces associated with sugar. This process can be readily traced in England, where members of Parliament were often elected by 'rotten boroughs', localities which might once have had many voters but now had a bare handful who did as they were told.

All forms of sugar were highly taxed, and the English government (amongst others) took the view that as much profit as possible should be returned to the 'home' nation. This meant that the sugar colonies were forced to send raw sugar back to Europe, where the value-added procedures of refining could be carried out. As those who are taxed always do, the planters complained, and sought relief from the taxes, which they saw as adding to the price—or worse, filching their profits. Not unnaturally, they felt that all the money paid by the end consumer should go to them, the sugar producers. In England, because the planters were rich (even with the taxes they were paying), 'sugar interests' were able to buy up many of the rotten boroughs, in order to force changes favourable to them, by holding the balance of power.

In 1767 Lord Chesterfield tried to buy a rotten borough for his son, but found that the cupboard was bare, according to a borough-monger who said that 'there was no such thing as a borough to be had now; for that the rich East and West Indians had secured them all, at the rate of three thousand pounds at least; but many at four thousand; and two or three, that he knew, at five thousand'. In the longer term, this abuse of the rotten boroughs helped to build up the pressure for electoral reform, leading to a more democratic Britain. The connection with sugar and rum is small, but they did play a part.

Around 1780, King George III was riding in a carriage with the then Prime Minister, Pitt the Elder, when he saw a carriage far outclassing his. He asked who the owner was, and learned that it belonged to a West Indian planter. According to the tale, the King, who was not always dotty, said, 'Sugar, sugar, eh? All that sugar. How are the duties, eh, Pitt, how are the duties?' The tale may be apocryphal, but it illustrates the nature of sugar wealth, and how it was seen as a source of revenue.

This was the pattern of late eighteenth- and early nineteenth-century England that Jane Austen showed us in *Mansfield Park*, where a plantation-owning family had become immensely rich and bought themselves an estate. The 'new rich' were looked down on by their neighbours, but as they spent their way into power and influence, so the West Indian plantocracy managed to infiltrate the lower ranks of the aristocracy, and to become, at the least, suitable for marrying to younger sons and daughters.

There was a major difference between the French sugar islands—which were a market for French brandy—and the English islands—which provided rum for the English drinker. While Jamaica and the other English islands used all of their molasses to make rum for a market they controlled, the French planters needed to dispose of their molasses somewhere away from France, so quite a lot of it ended up in what were then Britain's American colonies.

Much French molasses was shipped to Rhode Island, where it was converted to rum and smuggled to the other colonies. Here we find one of the major causes of the American Revolution— but to place it properly, we need to consider the Seven Years' War of 1756–1763, mainly involving Britain and France in a struggle for world supremacy. It was a worldwide conflict fought in Europe, North America and India, with Austria, Russia, Saxony, Sweden and Spain joining with the French, and Prussia and Hanover supporting Britain. It ended with the defeat of the French.

In the peace negotiations that followed, neither side cared much about who ended up with the Canadian colonies, but the sugar colonies were valuable. There were those in England who thought in terms of tax revenues, and argued that Britain should hold on to the captured French islands of Martinique, Guadeloupe and St Lucia. Against this, England's 'American interests' wanted the French out of North America. The sugar interests realised that sugar from a British-owned Guadeloupe would compete in the English markets on equal terms with their own sugar, and they argued for the return of the Caribbean islands to France. In the end, England offered an exchange: some of the sugar islands for all of Canada (an offer which Voltaire mocked as exchanging sugar for snow).

This exchange had far-reaching consequences, but it was only part of the war settlement. To prevent the entire Louisiana territory falling to the British, in 1762 the French had secretly ceded (under the Treaty of Fontainebleau) the area west of the Mississippi and the Isle of Orleans to Spain. The Treaty of Paris (the 1763 peace treaty) ceded all French territory east of the Mississippi, except the Isle of Orleans, to the British, which meant that it would soon become part of the new United States. In an 1800 treaty, Spain transferred the land west of the Mississippi back to France, setting the scene for the Louisiana Purchase. The cards were dealt, the play was about to begin—and once again, the play revolved around sugar.

Returning to the Seven Years' War: after 1763, Britain continued to hold the 'ceded islands' of Dominica, Grenada, St Vincent and Tobago, which were still allowed to import timber from Canada and the New England colonies, and to export molasses and rum to those places. The next year England made an unwise move, when Parliament passed the *Sugar Act* of 1764. The effect of this was to continue the *Molasses Act* of 1733 and place a duty

> . . . for and upon all white or clayed sugars of the produce or manufac-
> ture of any colony or plantation in America, not under the dominion of
> his Majesty . . .; for and upon indico, and coffee of foreign produce or
> manufacture; for and upon all wines (except French wine) for and upon
> all wrought silks, bengals, and stuffs, mixed with silk or herba, of the
> manufacture of Persia, China, or East India, and all callico painted,
> dyed, printed, or stained there; and for and upon all foreign linen cloth
> called Cambrick and French Lawns, which shall be imported or
> brought into any colony or plantation in America, which now is, or
> hereafter may be, under the dominion of his Majesty . . .

Other parts of the Act placed duties on rum, spirits and molasses, and the Americans suddenly realised that England was surplus to their requirements. The merchants of Rhode Island had lost a great deal of shipping in the Seven Years' War and now were suffering from a tax that cut directly into their profits—and they did not like it. The smugglers of Rhode Island also found themselves under threat, and in 1764 the excise schooner *St John* was driven off by fire from a shore battery—Rhode Islanders still claim these as the first shots in the coming war. Another excise schooner, the *Liberty*, was burned by the people of Newport in 1769. In 1772 a third schooner, the *Gaspee*, was lured onto a sandbar and the Rhode Islanders went out in long-boats and burned the ship.

Britain had been able to get away with taxing the American colonies when the French in Canada were a threat, but now, because of Britain's own choices under the Treaty of Paris, the colonies no longer felt at risk, and concluded that they were free to control their own destiny. It was sugar politics that had cleared the major part of the North American continent of other European powers; and it was sugar politics that led to the formation of an independent nation that could buy the land west of the Mississippi from the French, and so extend the realm of the English speakers across the whole continent. That in turn allowed the expansion of the United States, so early nineteenth-century sugar politics and policies can be said to have played a very real part in the shaping of the balance of power in the twentieth century.

On a much smaller scale sugar also shaped the future of Fiji. While formal power had been ceded to the British by the islands' chiefs in 1874, the Governor, Sir Arthur Gordon, was determined to avoid the land alienation that had happened in Hawaii, Australia and New Zealand. To make sure planters and other immigrants did not take over all the land, he formalised the Bose Levu Vakaturaga, the Great Council of Chiefs, to advise the governor on 'Fijian affairs'. The tensions now seen between the descendants of Indian indentured labour and modern Fijians stem from this decision.

SEA POWER, SUGAR AND WAR

One curious effect of the American War of Independence was the British build-up of naval power. After a fleet of 24 French ships stopped a British fleet from entering Chesapeake Bay to relieve the army of Lord Cornwallis in September 1781, Britain's navy went into a decline which saw most of the British sugar colonies picked off by the French until Admiral Rodney's victory at The Saints in 1782 stopped the French taking Jamaica, the greatest prize of all. Whatever interpretation we put on the French role in America's war, the French had no doubts of what it was really about: the control of sugar colonies and sugar incomes.

After the peace of 1783, Britain began to rearm, building 43 new ships of the line, repairing 85 others, establishing a base in Australia and another on Norfolk Island in the nearby Pacific Ocean. (The French had looked at the island and dismissed it as only 'fit for angels and eagles to reside', because it lacked safe anchorages.) While Norfolk Island was useless as a naval base, the British saw the Norfolk Island pines as timber for masts and spars, and they set about growing flax as the raw material for sail canvas. They 'imported' (that is, kidnapped) two Maori men to weave the flax, not realising that among the Maori, weaving was done only by the women.

Most Australians believe their nation was created solely as a dumping ground for convicts, but the authorities also saw it as establishing a base that might support the British navy in the East Indies, still very much a theatre of war. James Matra, a Loyalist who had sailed to the Pacific and visited Norfolk Island and the Australian coast and Botany Bay with Cook in 1770, argued in 1783 for an Australian settlement, drawing in timber from New Zealand for naval purposes, and growing spices:

> . . . as part of New South Wales lies in the same latitude with the Moluccas, and is very close to them, there is every reason to suppose that what nature has so bountifully bestowed on the small islands may also be found on the larger. But if . . . it should not be so . . . as the seeds are procured without difficulty, any quantity may speedily be cultivated.

Matra argued that 'we might very powerfully annoy' either Holland or Spain from the new base in time of war. He also argued from time to time that Australia might offer a good home to American loyalists, and that various trades might prosper there, but oddly enough he never mentioned sugar, perhaps because Britain already had an excellent source of sugar in the West Indies.

It was this wealth of sugar, and the wealth of the sugar growers, who held a bloc of seats in the British Parliament, that would shape British naval policy and rearmament and influence the way in which the 1793 war with France (which ran in fits and starts until 1815) would be fought. In the long term, the far greater maritime and naval strength of Britain in the Caribbean could be seen to have encouraged sugar growing in other areas, particularly Mauritius and Brazil—and, in the longer term, to have fostered the use of sugar cane's greatest competitor, the sugar beet.

BRANDY FROM OIL COGNAC

Take of pure spirits 10 gallons, New England rum 2 quarts, or Jamaica rum 1 quart, and oil cognac from 30 to 40 drops, put in half a pint of alcohol, colour with tincture of kino, or burned sugar, which is generally preferred. Mix well and bottle.

Daniel Young, *Young's Demonstrative Translation of Scientific Secrets*, Toronto, 1861

8

THE END OF SLAVERY IN THE AMERICAS

It is estimated that between 1450 and 1900 around 11.7 million slaves were exported from Africa, their destination the Americas. Only 9.8 million reached the other side of the Atlantic, however, the difference representing the numbers who died in the Middle Passage. To that we must add those killed or maimed in slave-taking raids, those who died while waiting to be brought to the coast, those who died while waiting to be sold, and the children who died because their parents were taken. If we then add to that figure the numbers who died as an indirect result of the slave trade, the victims must surely have numbered close to 20 million. The vast majority were put to work growing sugar, by far the worst of the operations a slave might be engaged in, once the Spanish mines stopped yielding riches.

In the eighteenth century, the mills were able to extract only about half the juice from the cane, which meant a plantation had to produce 20 tons of cane to obtain one ton of sugar. As well, it was necessary to haul away the bagasse (cane waste), fetch firewood, at the rate of five tons per ton of sugar, and take down the vanes on the windmills in the hurricane months. Across the world it was considered a fact that white men could not work at physical tasks in those conditions where sugar cane thrived. Black people were needed for that—slaves, in other words. Names, though, were clearly important, as 'No Planter' made clear in a letter to the *Gentleman's Magazine* in 1789:

> The vulgar are influenced by names and titles. Instead of SLAVES, let the Negroes be called ASSISTANT-PLANTERS, and we shall not then hear such violent outcries against the slave-trade by pious divines, tender-hearted poetesses, and short-sighted politicians.

'No Planter' meant the common herd when he referred to the 'vulgar', but most of those in the sugar islands were vulgar adventurers. Planters went to the colonies only for the time it took to make their fortune—as Richard Ligon noted during his tour of Barbados: 'Colonel Thomas Modyford has often told me that he had taken a Resolution to himselfe, not to set his face for England till he made his voyage and employment there worth him a hundred thousand pounds Sterling, and all by this sugar plant.'

In fact Modiford, to give him his more usual spelling, became Governor of Jamaica in 1664, where he and the planters who accompanied him from Barbados introduced sugar cultivation. Modiford was recalled to Britain after he and Morgan declared war on the Spanish, and committed to the Tower of London, but in 1675 he was allowed to return to Jamaica and died there in 1679. That was not part of his original plan.

TRADE AND WELLBEING

The simple fact was that sugar and slavery had served too well the needs of those who made the decisions and wrote the rules, and that made slavery hard to bring to an end. In 1689, Edward Littleton published a pamphlet entitled *The Groans of the Plantations*. From the title this might have been assumed to be about the plight of the slaves; instead, it dealt with the needs of the poor, suffering planters, complaining that a wicked, plundering, profiteering company stood between them and buying slaves cheaply on the Guinea coast.

> Of all things we have occasion for, *Negroes* are the most necessary and the most valuable. And therefore to have them under a Company, and under a Monopoly, whereby their Prices are doubled; cannot be but most grievous to us . . .
>
> In *Barbados*, we can get but little by making Sugar (though it had none of these Burdens) except we improve it: that is, purge it and give it a Colour. Others can live by making plain Sugar: We must live by the improved . . . But the Duties fall so terribly upon our improved Sugars, that it doth quite discourage and confound us.

But it would be wrong to think that Littleton did not care for the needs of the slaves as well:

> In the mean time our poor Slaves bear us Company in our Moans and groan under the burden of these heavy Impositions. They know that by reason of them, they must fare and work the harder. And that their Masters cannot now allow them, and provide for them, as they should and would . . .
>
> We must have yearly some hundred pairs of Sugar-Pots and Jars. Every hundred pair doth cost near ten pound; and we must fetch them several Miles upon *Negroes* heads.

In other words, if you care about the poor black slave, afford the master the opportunity to make more money and all will be well for the slave. The outcry against taxes revealed the nature of the tensions in the colonies, with the home government assuming that the colonies, founded with capital provided by the home nation, should exist only to enrich those at home and the home government. The fear of a slave revolt also grew as the slaves began to outnumber whites. In the later eighteenth century, the Jamaican House of Assembly wanted to reduce the number of slaves coming in by imposing a duty, but this was refused in 1774, and when they protested were told from England: 'We cannot allow the colonies to check or discourage in any degree a traffic so beneficial to the nation.'

This attitude to the colonies lasted: indeed, it was the main factor in alienating the thirteen American colonies. It was not peculiar to England, however—the home governments made the rules, and they all made the rules to favour the trade and well-being of the home nation, no matter which nation it was—or which party was in power. An Act of the English Parliament placed a high import tax on refined sugar in 1685, ending refining in the colonies. As the French had fewer refineries, they originally encouraged colonial refining in order to cut down on imports from the rest of Europe, but they barred colonial refining as soon as they had refineries of their own, thus transferring the profits from this activity to the home nation.

England's *Plantations Duties Act* of 1673 provided that goods taken from one colony to another would be charged the same duties they would have been charged coming into England, but that was only as effective as its enforcement, something Mitford Crowe, Governor of Barbados, knew when he wrote in 1707:

> I must inform Your Lordships, it would be some help to this Island if the trade between New England and Surinam were obstructed, for if I bee rightly informed great quantity of Rum, Sugar and Molasses go in return for their horses, flower and provisions, and last week one Harrison, a Planter, being much in debt, run off in a sloop with sixty negroes leaving his land to his Creditors.

The slave traders could also make financial losses. The Middle Passage, the route from Africa to the Indies, was fraught with risk from storms and from attack on the African coast, and from the loss of slaves to disease or despair. However, the trade from the West Indies to Europe was generally profitable, and the European goods sold to the Africans were so cheap and shoddy that it was almost impossible to lose out. The poor quality of goods made for Africa was such that when Arthur Phillip wished to complain about the tools that had been supplied to his settlement at Botany Bay, he wrote: 'I cannot help repeating that most of the tools were as bad as ever were sent out for barter on the coast of Guinea.'

Risky or not, the triangular trade made many people rich. Lewis Carroll, baptised Charles Lutwidge Dodgson, had a middle name that honoured his great-grandfather, a slave trader. Edward Gibbon could afford to write his histories because his grandfather had been a director of the South Sea Company, a slave carrier. The Vicomte de Chateaubriand's father was a slave captain, and later a slave merchant, but nobody thought the less of Chateaubriand as a liberal and a man of letters. John Locke, the philosopher, was a shareholder in the Royal African Company, another slaving concern, even though he wrote: 'Slavery is so vile and miserable a state of man, and so directly opposite to the generous temper and courage of our nation, that it is hardly possible that an Englishman, much less a gentleman, should plead for it.'

In Denmark, the dominant Schimmelmann family was also wealthy, thanks to slaving—and provided various governments with two ministers of finance. Members of the British Parliament represented the West Indies and sugar interests well, and sound men of trade, Lord Mayors of London among them, were all advantaged by the trade in human flesh.

THE ZONG CASE

Much evil took place in the slave trade, and the *Zong* case, often called, with reason, the *Zong* atrocity, is perhaps the best example of this. In November 1781, Luke Collingwood, master of the *Zong*, threw 132 slaves overboard to drown and then claimed insurance on them. He still had 440 slaves on board, and claimed that a lack of water had forced him to jettison and drown the others in order to save the ship, the remnant human cargo, and the crew.

Not surprisingly the insurance company objected—but not on humanitarian grounds. If the slaves had died natural deaths, the insurers argued, there could have been no claim. The abandoned 'property' had been sick and dying, and this was an attempt to exploit a legal loophole. There was no shortage of water: neither crew nor slaves had been on short rations and the ship arrived at Jamaica on 22 December with 420 gallons of water remaining. In that case, the captain had not acted out of necessity, and could not claim £30 a head on the drowned slaves.

The Solicitor-General, John Lee, spoke for the owners and made the official and legal view of slaves clear, when he asked in court:

> What is all this vast declamation of human people being thrown over-board? The question after all is, was it voluntary or an act of necessity? This is a case of chattels and goods. It is really so: it is the case of throwing over goods; for to this purpose, and to the purpose of the insurance, they are goods and property: whether right or wrong, we have nothing to do with it.

The owners won their claim for insurance, but it was later over-turned on appeal. The only clear loser was the slave trade, for this callous relegation of humans to the status of chattels ignited a fire that would not be easily extinguished. Collingwood and his men were never tried for murder, but the insurance trial sparked the first real calls for an end to the trade in human lives, whatever the effect that might have on the sugar trade—and that worried some people.

ABOLITION AND EMANCIPATION

The great British reformer William Wilberforce was barely into his teens when the cause of emancipation had a great victory. Lord Chief Justice Mansfield said in 1772: 'The status of slavery is so odious that nothing can be suffered to support it but positive law.' Since there was no such law, he declared that a slave setting foot in England was thereby a free man or woman. All the same, a motion to oppose slavery was defeated in the House of Commons in 1776. While this remained the pattern for 30 years, emancipation slowly gained ground.

In May 1789, a year after Wilberforce entered Parliament, Richard Pennant told the House that if they passed the vote for abolition as proposed by Mr Wilberforce, 'they actually struck at seventy millions of property, they ruined the colonies, and by destroying an essential nursery of seamen, gave up dominance of the sea at a single glance'. The opponents of slavery were no radicals, however, but men of serious demeanour and sound business principles.

There were other ways of training sailors than by sending them on slaving ships, and other ways of making money than by trading in slaves and slave goods. Many good men opposed emancipation in the name of profit, but that same wholesome pursuit of wealth brought some less than altruistic support for the emancipation movement from interests involved in the East

Indies trade, which was by now producing sugar in India, with hired rather than slave labour. By 1792 the East India Company was claiming that the cost of a West Indies sugar slave's life was 450 pounds of sugar. It was said, in tones of moral outrage, that 'a family that uses 5 pounds of sugar a week will kill a slave every 21 months'. To rub the message in, the company told consumers that eight such families, in just nineteen and a half years, would kill 100 slaves with their sugar consumption!

William Fox was the first to raise the spectre of dead slaves against West Indies sugar. Fox also knew how important rum was to the sugar industry, and in 1781 he was quite clear in calling for people to abstain from both rum and sugar:

> . . . until our West Indian planters themselves have prohibited the importation of additional slaves, and commenced as speedy and effectual subversion of slavery in their islands as the circumstances and situation of the slaves will admit; or until we can obtain the produce of the sugar cane in some other method unconnected with slavery and unpolluted with blood . . . a family that uses 5 lb. of sugar per week with the proportion of rum will, by abstaining from the consumption for 21 months, prevent the slavery or murder of one fellow creature.

In other words, 450 pounds of sugar, and the amount of rum made from the molasses remaining after making that sugar, was the equivalent of the sugar produced by one slave over the 10-year life expectancy that a sugar slave had *after reaching the cane fields*. Other campaigners gave the weight of sugar as 405 pounds, but both sides were guilty of massaging their figures.

In fact, the 'exchange rate' in the plantations was one ton of sugar per slave life in 1700, and two tons per slave life in 1800. The amount of sugar that today's average high school population in the developed world consumes in one week, as junk food, confectionery, ice cream and soft drinks, is enough to have killed a slave in 1800. The exchange rate was a very powerful statistic, capable of being tailored to almost any audience.

As part of its promotions, the East India Company distributed sugar bowls with the legend 'East India Sugar not made by SLAVES'. Their advertising pamphlet used the same statistics on slave deaths, and East Indies sugar was sold at a considerable premium—while West Indies sugar sold from 70s. to 80s. per cwt around 1792, East India sugar was selling at 140s. The curious logic of the pamphleteers allowed the West Indies faction to use this to argue that there should be no excise break for East Indian slave-free sugar, since clearly people were willing to pay that sort of premium for slave-free sugar!

One of the first real breaks for the slaves came at sea, off Trafalgar in 1805, when Nelson broke French sea power once and for all. Now Britannia ruled the waves, and what Britannia said, others did, or else—and Britain said it was time to stop shipping slaves. The war had stopped all the world's slaving ships sailing for some years, and planters had been forced to look after their slaves better, knowing there were no replacements. Clearly, the practical men of Parliament said, it would be a good thing to ban the trade forever, so the slaves would continue to be treated with kindness.

It would be a better thing to stop the trade now, they said, because there would be no slave traders demanding compensation for the ending of their trade. Denmark, a minor player in the trade, had already banned the slave trade, and so had the United States in 1800, when American ships were forbidden to carry slaves to foreign countries. That was extended when the slave trade in general was abolished by the US Congress just three weeks before the British bill was given assent by the king. That left only France, which had no sea power in any case. But knowing that the West Indies faction had enough members to make and break government majorities, the government did not free the West Indies slaves. England needed slavery—or, at least, the British government needed the votes of the British MPs controlled by the West Indies slave owners.

By the turn of the nineteenth century, the new-breed entrepreneurs were moving into canals and turnpikes—railways were still some time away but steam engines were becoming common. The first steam-powered mill in Jamaica was as early as 1768, but now steam was powering breweries in London, mines in Cornwall and many other ventures. England's rich and powerful were finding new investments, and new ways of controlling labour.

In 1799, even as he was working to free the slaves on the plantations in the sugar colonies, William Wilberforce obtained the passage of a bill that forbade working people in England from forming combinations to defend themselves against the tyranny of the masters, thus reducing miners, textile workers and agricultural labourers to the status and conditions of industrial slaves. He seemed not to see the inconsistency of his position.

Over the next ten years, a number of European powers limited or renounced slavery, and slowly the trade was rolled back. In 1816 the slave owners of Ceylon voted voluntarily that the children of slaves born after 12 August in that year would be free born.

While slavery remained in force, the Consolidated Slave Law enacted in Jamaica in 1816 improved living conditions somewhat, and we can take this as a good indication of the practices applied on the other British sugar islands. The law set down that slaves should have Sundays off and one other day off each fortnight to work on their own produce grounds, except during harvest time. The minimum of free days was not to be less than

26 per year, and there was to be no milling of cane between 7 pm Saturday and 5 am Monday. Field work was restricted to the period from 5 am to 7 pm, with half an hour for breakfast, and a two-hour break in the middle of the day.

In 1823 the emancipation movement was renamed as the 'Society for the Mitigation and Gradual Abolition of Slavery throughout the British Dominions'. There were still pockets of slaving, and the various powers took umbrage at their ships being stopped and searched by naval vessels from foreign anti-slaving nations. In any case, a slave ship could always be 'foreign', presenting different papers as necessary. Unless several nations' navies were to operate together, the Middle Passage would still be sailed, and more 'black ivory' would be delivered to the Americas. Slaves even reached the United States, where slaves were landed for many more years in the non-state of Texas, and marched across the country.

Most of the opponents of slavery recognised that it still existed, but the state of slavery, they believed, was less harmful than the trade in slaves. Besides, if the slave trade was stopped, those few bad masters would have to take better care of their slaves, because the supply of replacements had dried up when the slave ships were stopped from sailing. Like modern economic rationalists, they saw market forces as the best way of doing good.

It is worth noting that for all their limitations, the dissenters and evangelicals did one thing for slaves that nobody had done before: they welcomed them as Christians. In the early days of sugar slavery in the Mediterranean, slaves controlled by Muslims could not enter a mosque, and Christian-owned slaves were generally unwelcome in a church—even though Muslim slaves built many churches and Christian slaves built many of the mosques.

In the more civilised and genteel climate of the nineteenth century, Baptists and Wesleyans moved in to preach to their black brothers and sisters, who might still be slaves but whose souls would eventually be set free. This gave the planters the cheap labour they needed, while satisfying the urge of the missionaries to give their charges the promise of a better afterlife. Of course, Edward John Eyre, who found most of the unrest coming from within the Baptist churches, may have had a different view.

CHRISTMAS CAKE

5 teacupfuls of flour, 1 teacupful of melted butter, 1 teacupful of cream, 1 teacupful of treacle, 1 teacupful of moist sugar, 2 eggs, ½ oz. of powdered ginger, ½ lb. of raisins, 1 teaspoonful of carbonate of soda, 1 tablespoonful of vinegar. Make the butter sufficiently warm to melt it, but do not allow it to oil [sic]; put the flour into a basin; add to it the sugar, ginger, and raisins, which should be stoned and cut into small pieces. When these dry ingredients are thoroughly mixed, stir in the butter, cream, treacle, and well-whisked eggs, and beat the mixture for a few minutes. Dissolve the soda in the vinegar, add it to the dough, and be particular that these latter ingredients are well incorporated with the others; put the cake into a buttered mould or tin, place it in a moderate oven immediately, and bake it from 1¾ to 2¼ hours.

Mrs Beeton's recipe number 1754

9
EMANCIPATION'S HARVEST

Slavery was finally abolished in the British Empire in 1833, through a deal in which slave owners throughout the Empire received around £20 million in compensation, and were allowed to keep their slaves for an 'apprenticeship' period of twelve years. This 'apprenticeship' turned out to be a sham, and most owners made no attempt to use it as a period of transition leading to freedom, instead setting out to ensure that their slaves remained dependent. The slaves, understandably, could see no reason why they should not be freed immediately.

In 1837 a revised deal was cut, involving the termination of 'apprenticeship', immediate liberation and the imposition of a tariff on sugar not produced in British possessions. Even that did not satisfy the slave owners, who found many ways of using arbitrary local laws and high rents to force the freed slaves back into effective slavery. There were occasional burnings of black settlements that were 'too distant' from the plantations, aimed at bringing the slaves back under tight control. The English were doing exactly the same thing in Ireland at the same time, something that Thomas Carlyle obviously had in mind when he likened Ireland and the sugar islands:

> Our own white or sallow Ireland, sluttishly starving from age to age on its act-of-parliament 'freedom,' was hitherto the flower of mismanagement among the nations; but what will this be to a Negro Ireland, with pumpkins themselves fallen scarce like potatoes?

This and the other selections quoted in the next few pages are taken from Carlyle's pamphlet, first published in 1849 and revised in 1853 (the selections are from the 1849 original). It came with the unfortunate title *Occasional Discourse on the Nigger Question*. Carlyle was setting out to be offensive, but his title was far and away the least of his offences.

Assisted in part by the British and Foreign Anti-Slavery Society (the BFASS, referred to by Carlyle as 'Exeter Hall', after the place where the emancipists met), the freed slaves began a mass exodus from the plantations to 'free villages' and their own small farms in the hills of Jamaica and other West Indian islands. Settled on their 'pumpkin' farms, as Carlyle called them (he meant melons when he wrote pumpkins), they became subsistence farmers, creating a sudden labour shortage that was only partly remedied by the introduction of Indian indentured labourers. Soon, though, things got worse.

TAXATION POLICY

At the same time as the slaves were 'freed', Britain's Parliament, having repealed the Corn Laws, which lowered the price of the workers' bread, decided also to reduce the price of the workers' sugar, because by now, everybody needed sugar. To this end Parliament equalised all sugar duties in 1845. Cuban sugar—still grown by slaves—entered Britain at the same duty as sugar grown by free workers. The plantation owners in the British West Indies knew their sugar could not compete with the slave-grown products of Cuba and Brazil. As they saw it, they had been betrayed.

In two other British colonies on the other side of the world, duties and tariffs imposed by London were also to cause distress a generation later. English novelist Anthony Trollope noted how the potential for a preserved fruit industry in Tasmania in the 1870s was blocked by the requirement that sugar from Queensland, another Australian colony, be taxed as though it came from a foreign land, taking away the opportunity for profitable commerce in both colonies.

Taxation policy seems to have been a problem for successive English governments. They were slow to learn from the harm they had done to both their American colonies and themselves by their taxes and duties, and continued to make laws for their colonies that did active harm (aside from giving Cuban slave-grown sugar some handy assistance).

Curiously, the Cuban sugar trade had been created by England, whose military forces occupied the island in 1762, towards the end of the Seven Years' War, so opening it to English trade. More than 10 000 slaves were brought in, some say in anticipation of England holding the island thereafter. Others believe that they were brought in as porters for the army, though the army might also have had in mind that a handsome profit might later be made on their 'army surplus'. Whatever the truth of the matter, Cuba was returned to Spain, the slaves remained on Cuba, and by 1865 as many as 2 million more had been landed there.

The BFASS was at first satisfied with the new indentured labour system, the system of 'apprenticeship'—which was to be condemned in the Pacific Islands by the London Missionary Society 40 years later as poorly disguised slavery. The BFASS tacticians in London felt that if free (though indentured) workers could be shown to grow sugar successfully, then slavery in the United States might also be abolished. Exploitation was a lesser evil than slavery, they seem to have thought, though this belief later changed.

At the same time, a battle was being fought between Free Trade and Protectionism. The laws of supply and demand, the notions that we would today call economics, the notions Carlyle in his pamphlet called 'the dismal science', won out. Sugar and slavery were so interconnected that any support for sugar would also be support for slavery, the Free Traders said. This rather specious claim took hold and led to Cuban sugar being allowed into Britain on the same tariff as sugar grown by freed workers. Soon after, the British Caribbean sugar industry went into a decline, and Carlyle's indignation soared. He wrote of the ungrateful blacks:

> . . . with their beautiful muzzles up to the ears in pumpkins, imbibing sweet pulps and juices; the grinder and incisor teeth ready for every new work, and the pumpkins cheap as grass in those rich climates, while the sugar-crops rot round them uncut, because labour cannot be hired, so cheap are the pumpkins; and at home we are required but to rasp from the breakfast loaves of our own English labourers some slight 'differential sugar-duties,' and lend a poor half million, or a few poor millions now and then, to keep that beautiful state of matters . . .

The economic decline affected all whites in the West Indies—not only the plantation owners, but also the associated clerks, foremen and others who gained their income from sugar. The freed slaves, happy to live a subsistence lifestyle, had no problems, Carlyle wrote:

> . . . the less fortunate white man of those tropical localities . . . himself cannot work; and his black neighbour, rich in pumpkin, is in no haste to help him. Sunk to the ears in pumpkin, imbibing saccharine juices, and much at his ease in the creation, he can listen to the less fortunate white man's 'demand,' and take his own time in supplying it. Higher

wages, massa; higher, for your cane-crop cannot wait; still higher, till no conceivable opulence of cane-crop will cover such wages.

This, of course, was tied to the much-repeated claim that people of European origin could not work in the tropics—even though they had done so on Barbados, 200 years earlier, as indentured labourers. (The same line was later used to justify the Australian Kanaka trade.) Carlyle's main assumption was that the feudal form of life was far more satisfying for all. While he might be seen as advocating a return to slavery, he really favoured serfdom, as these two excerpts reveal:

If quashee [the black man] will not honestly aid in bringing out those sugars, cinnamons, and nobler products of the West India Islands, for the benefit of all mankind, then I say neither will the powers permit quashee to continue growing pumpkins there for his own lazy benefit; but will sheer him out, by and by, like a lazy gourd overshadowing rich ground; him and all that partake with him—perhaps in a very terrible manner.

Lest that was not clear enough, Carlyle goes back over his case:

Not a square inch of soil in these fruitful isles, purchased by British blood, shall any black man hold to grow pumpkins for him, except on terms that are fair towards Britain. Fair; see that they be not unfair, not toward ourselves, and still more not towards him. For injustice is for ever accursed: and precisely our unfairness towards the enslaved black man has, by inevitable revulsion and fated revulsion of the wheel, brought about these present confusions. Fair towards Britain it will be, that quashee give work for privilege to grow pumpkins. Not a pumpkin, quashee, not a square yard of soil, till you agree to do the State so many days of service. Annually that soil will grow you pumpkins; but annually also without fail, shall you, for the owner thereof do your appointed days of labour. The State has plenty of waste soil; but the State will religiously give you none of it on other terms. The State wants sugar from these

> islands, and means to have it; wants virtuous industry in these islands,
> and must have it. The State demands of you such service as will bring
> these results, this latter result, which includes all.

In other words, work makes free, as a later economic rationalist once said. Carlyle was demonstrating a very feudal and European attitude to land—it was the property of the rich, the rulers; if others wanted to work the land, they could not do so freely, but must pay their betters for the privilege. The English and other Europeans came to the Caribbean as rulers, seized the land, and worked both the land and its occupants as hard as possible so that they could gain a huge profit, return 'home' and buy land, for 'home' landholdings would allow them to hold their heads high in the polite and genteel society of other landholders while grinding down their own agricultural poor.

Many British administrators saw the freed slaves as Carlyle did, when they thought of them at all, but not everybody took that view. The poet Matthew Arnold, for example, called Carlyle a 'moral desperado'. Carlyle's friendship with John Stuart Mill had survived a disaster when Mill's maid mistakenly burnt the only manuscript copy of Carlyle's *The French Revolution*, after which Carlyle rewrote the entire draft from memory, but it did not survive the response that Mill made to Carlyle's pamphlet on the Negro question. Mill was in favour of social reform based on his utilitarian philosophy; he saw labour as a necessary evil, while Carlyle saw it as a virtue in itself, a duty that the black must attend to, forthwith, for the general good.

THE EYRE CASE

Mill and Carlyle were destined to clash again, as a result of a minor scuffle at Morant Bay in Jamaica in 1865. Reported to the Governor of Jamaica, one Edward John Eyre, as a 'black uprising', it led him to act fast. Eyre was the English teenager we met earlier, sipping tea and rum in the Australian bush. He is still well known to Australians for his exploration of the continent between 1836 and 1841, and his name is found on the great salt pan, occasionally a body of water, called Lake Eyre.

After walking half across Australia, Eyre became Protector of the Aborigines, distinguishing himself mainly by his attempts to convert the Aboriginal people of South Australia into dark-skinned Britishers, asking them to give up their culture for that of the white man. In the process, he risked his own life many times to save others, intervening to stop fights between armed white stockmen and armed Aboriginals. By that time, white settlers were moving across Australia from Sydney to Adelaide, with large flocks and herds, fighting battles with the local people whose land they crossed. As usual, more blacks than whites died in these encounters.

Seeking to rise in the world, Eyre went off to New Zealand as Lieutenant-Governor. There, he was treated inhumanely by his superior, and is now recalled mainly for his obsession with gold braid, ceremonial matters and Sabbath observance. He then moved to the West Indies, becoming in turn Lieutenant-Governor of St Vincent, and then Antigua, before being appointed Governor of Jamaica in 1864.

When whites were killed at Morant Bay, Eyre acted as ruth-lessly as the worst of the white herders and settlers he had tried to control in South Australia. He declared martial law and his forces hanged more than 400 blacks. While this firmness of resolve was applauded by the local white community—and even some of the black community—Eyre made one bad mistake: he hanged a prominent political opponent, a man of colour named

George William Gordon, without a fair trial. To make it worse, the hanging took place outside the area declared under martial law. Eyre followed up this action by abolishing the Jamaican legislature and constitution.

There was uproar in Britain. Karl Marx reacted to the events by likening the beastliness of the 'true Englishman' to that of the Russian. Eyre was recalled to London, and many famous Englishmen, including Mill, Charles Darwin, Thomas Huxley and Herbert Spencer, called for him to be charged with murder. The conservative forces, led by Carlyle, Alfred, Lord Tennyson and John Ruskin, created the Eyre Testimonial and Defence Fund. In the end, Eyre was allowed to die peacefully in Devon in 1901, looking and behaving more like the gentle and humane Protector of Aborigines than the Hangman of Jamaica.

COMPETITION FROM SUGAR BEET

By the 1860s more and more countries were growing sugar cane, partly for their own use, but also for export, and that pushed prices down. Even more of a problem was the spread of sugar beet, which could be grown in temperate regions all over Europe, the biggest market. By now the art of getting fine sugar crystals from a sugary solution was common knowledge, because refining had always been reserved as a 'home' industry.

The wonder of sugar beet is not so much that people found a way to grow sugar in temperate conditions, but that they took so long to adopt an ancient discovery. Theophrastus, who died in 327 BC, wrote that the white beet was richer in juice than the black, and cooks certainly knew it as a sweet-tasting root. Like sugar, the sugar beet was seen as having medicinal values. Avicenna (980–1037) recommended using the beet against nose and throat infections. Around 1575, Olivier de Serres (1539–1619) wrote that the cooked beet yielded a juice like sugar syrup, although he does not appear to have suggested using it as a source of solid sugar.

By the early eighteenth century, the development of the scientific tradition was well under way, and if there were yet few scientific journals to disseminate discoveries, at least there was an informative publication known as a magazine, and others would follow. This 1731 British invention took its name from the Arabic word *makhazan*, meaning 'storehouse' or 'warehouse', and magazines were seen as storehouses of information. For a century and a half, one such 'storehouse', the *Gentleman's Magazine*, spread ideas and opinions, and provoked readers to think and inquire. It is a remarkable source of information on the sugar trade—and on beet sugar, though its reliability was occasionally a little off.

THE SUGAR BEET

In 1747 Andraeas Sigismond Marggraf (1709–1782) reported that the sugar in sugar beet is identical with that found in cane. Marggraf was an eminent chemist who made many other discoveries, but today he is recalled for showing that the sugar in the root of the beet was the same as cane sugar.

Of course, not everybody was prepared to allow him due credit. In 1752 the *Gentleman's Magazine* had recorded a plan to extract sugar from seaweed, and even though Marggraf had done his work five years earlier, the magazine claimed part credit in 1754, when they heard of it. Here it is, with original spelling and punctuation, from the *Gentleman's Magazine*, 1754, No. 24, page 9:

We are pleased to find, that the account we gave in our *Mag.* for *July 1752*, p. 324 of a method of extracting sugar from the broad-leav'd alga, a seaweed, has excited the curiosity of the Literati in foreign nations, to persue this discovery still farther; by examining more closely the essential properties of other plants. M. Marggraf, of the academy of Sciences at Berlin, has published the result of the experiments he has made on this subject, by which it appears that many common herbs contain large proportions of sugars.

Still, the magazine's proprietors had the full details for their readers. Here is Marggraf, in their translation, after he has explained how he showed that sugar could be dissolved in strong brandy, strained through a cloth, and then formed into crystals again:

Having prepared the way by this experiment, I took the roots of white beets, and having cut them into small slices, I laid them by the fire to dry, taking care not to burn them: I then reduced them to a coarse powder, and laid it to dry a second time, because it is very apt to contract moisture: Whilst this coarse powder was yet warm, I put eight ounces of it into a glass vessel, and pour'd upon it 16 ounces of brandy, so strong that it fired canon-powder.

After boiling, straining the mash through cloth and waiting, he obtained a small amount of sugar from this and other roots:

By this method . . . I obtained from the three roots . . . the following quantities of sugar.
1. From half a pound of the root of white beets, half an ounce of pure sugar.
2. From half a pound of skirrets, an ounce and a half of pure sugar.
3. From half a pound of red beets, one ounce and a quarter of pure sugar.
It is evident from these experiments that lime water is not at all necessary to dry and thicken the sugar, as some pretend, since the sugar chrystalizes without it.

> Being thus assured that there was a real sugar in plants, I endeav-
> oured to find out a less expensive manner of extracting it . . .

Now the interesting point here is that while we recognise Marggraf as the discoverer of beet sugar, his trials had pointed another way, to the water parsnip or skirret as the best source. Significantly, the common English name skirret comes from the Dutch *suikerwortel*, meaning 'sugar root', so Marggraf was hardly making a major discovery here. Known to the botanists as *Sium sisarum*, but identified by Marggraf as *Sisarum dodonaei*, the skirret was intro-duced into Europe from China around 1548, and was soon hailed on all sides as 'the sweetest of roots'. This was the plant he used.

> I took a certain quantity of skirrets; I cut the roots, whilst fresh, and
> pounded them as small as possible in an iron mortar. I then put them
> into a linen bag and pressed out the juice . . . I poured water upon the
> roots remaining in the bag, and pressed them a second time. I . . . let it
> stand to settle in a cool place for 48 hours: In which time it became
> clear, and a mealy substance settled to the bottom . . .
>
> The first clarification being thus made, I put some whites of eggs
> to the juice, and boiled it in a brass pan, scumming it continually,
> till no further impurities appeared upon the surface: I then passed it
> thro' a linen cloth, and the liquor was as transparent as the clearest
> wine.

After further boiling to condense it to a syrup, Marggraf let the syrup stand in a warm place, and after six months, he found crystals on the sides of the glass vessel. Collecting the crystals, he drained them of syrup, blotted them, dissolved them in water, and strained the solution through cloth. He then used methods we should recognise, for he:

> . . . boil'd it to the consistence of a thick syrup; then put a little lime
> water to it, and boil'd it gently until it became ropy. I then took it off

the fire, and stirr'd it about until it cool'd and thickened a little; after which I poured it into well-burnt earthen vessels in the form of a cone, closed at the small end with a wooden stopper, which vessels I put into others that were deeper, and set them in a temperate place.

THE NAPOLEON GAMBIT

So with Marggraf claiming in Germany to have produced best Muscovado sugar, far from the sugar islands, and setting out the details for others to follow, why did it take two more generations for beet sugar to be produced commercially? The answer appears to be in two parts. Sugar could still be bought more cheaply from the tropics because the existing beet varieties demanded too much effort for the amount of sugar that resulted.

It was clear that the beets varied, and Marggraf later identified white beet as giving the highest yield, followed by skirret and red beet, but he recommended the beet as a source of syrups for cooking, not as the basis of an industry. According to the usual account, that was as far as the sugar beet went until the supply of cane sugar to Europe by sea was stopped by the Napoleonic wars. In fact, work was continuing behind the scenes. Marggraf's former pupil and successor in his chair, Carl Franz Achard, began a systematic study of beet sugar in 1786 at Caulsdorf. Thus, by the time Napoleon's many wars caused sugar shortages, the technology to produce beet sugar was available.

In 1799 Achard presented Frederick William III of Prussia with a loaf of sugar prepared at a Berlin refinery from raw beet. With regal assistance, Achard set up a sugar works. Unfortunately it failed, mainly due to his lack of business acumen, although insufficient research and development may have played a part as well. Achard also told the Institute of France of his results; but because the French were investigating getting sugar from grapes, nothing came of it then. Soon, though, France needed to find a replacement for the 100 000 tons of sugar that had come from Haiti each year, where the former slaves were no longer willing to carry out the back-breaking work required.

A Königsaal refinery in Bohemia produced beet sugar in 1797, and another beet sugar factory opened at Horowitz in 1800. Achard was by no means alone in his discovery, but it was the French who now took over, crossing various strains of beet and carrying out systematic tests of the new plants. The factory of Freiherr Moritz von Koppy started production in 1806; his 'White Silesian' beetroot has provided all of the modern strains of sugar beet. Achard also told the Tsar of his work, and a factory soon opened in Russia. By 1809 there were eight factories working there; three years later, there were eight factories around Magdeburg alone. Napoleon ordered an expansion of the French beet sugar enterprise, and by 1813 France had 334 factories producing close to 4000 tons of sugar.

Unfortunately, the bubble was about to burst. As Napoleon's armies retreated, Europe opened up again to English trade and to sugar from the English colonies. That started a long-lasting Gallic complaint of English sugar policy wiping out a delicate French industry, but the various tales of British attempts to kill off the sugar beet industry are at best dubious and at worst, total fabrications, probably having more to do with French domestic politics, and the need for a perfidious Britannic bogeyman, than to any planned English action.

If anything, the French sugar industry was doing better than the British. By 1815 France was at peace and subsidising her beet sugar industry, keeping beet sugar in competition with British cane sugar as prices plummeted. Having already lost most Caribbean sources of sugar, France also lost Mauritius in

1816, leaving little option but to develop beet sugar for domestic consumption. By 1826 some 1900 factories were making around 24 000 tons of sugar a year, and by 1833 consolidation into 400 factories had done no harm: production reached 40 000 tons of sugar annually, a third of the nation's needs. From there, it became a war of attrition between the two types of sugar.

GUNPOWDER

Take pulverized saltpetre, moisten it, and subject it to the action of a slow fire until completely dried and granulated, of this take 75 parts, purified sugar 12 and a-half parts, moisten and grind together till completely blended, which will require several hours, pulverize on heaters till dried.

Daniel Young, *Young's Demonstrative Translation of Scientific Secrets*, Toronto, 1861

10
THE RISE OF
TECHNOLOGY

M any problems faced the sugar cane planters of the Americas in the nineteenth century. They were beset by the loss of their slaves, by competition from new sugar-producing areas (both tropical cane and temperate beet) reducing the local demand and even competing for the export markets, and by irrational taxes and duties, all combining to reduce their profits.

They were left with no choice but to improve their performance. That meant applying the lessons of the Industrial Revolution and improving their crops by finding better canes and better ways to grow the canes. In the process, some people lost fortunes, others were thrown out of work and into grinding poverty, while a few made vast fortunes. The first step was to improve the technology—and that meant the end of the Jamaica Train, just as Brazilian growers were beginning to adopt it.

Sugar syrup boils at temperatures far higher than water boils, and one of the problems that Richard Ligon had noted around 1650 was that the great heat tended to burn out the copper pans in which the syrup was heated, especially as the level became lower. The last of the syrup would often 'catch', leaving a flavour and colour in the next batch of sugar. The cost of replacing burnt-out copper pans was high, but replacing them with cheaper iron pans increased the amount of 'catching', thus reducing the quality of the sugar, which was judged largely by colour.

Today, we would solve that by turning down the flame, but in those days, in the places where the sugar cane grew, gas was unavailable. The only real answer was to reduce the heat that was required, and that needed a vacuum pan, a sealed vessel under low pressure where syrup would boil at a sufficiently low temperature to stop the sugar catching.

An Englishman, the Hon. Edward Charles Howard, took out the first patent for a vacuum-pan method of evaporation in 1813, but it was a patent for a process, not a device. By 1827 there were just six of the systems in operation. The main aim was to concentrate the sugar solution while keeping inversion, the change to simple sugars, to a minimum. With its reduced temperature, the vacuum pan was a wonder, but it was largely an unused wonder.

Part of the problem was the taxation system then in place and the continuing attitude of the British government to sugar being refined in the colonies. The first vacuum-pan sugar arrived in England in 1833, and was deemed to be above the level of muscovado sugar, and so liable to a duty of £8 8s. per hundredweight. By 1845 it was accepted that this sugar should come in at 16s. 4d. per cwt, against 14s. for other sugars, the two classes being called 'equal to white clayed by any process' or 'yellow muscovado', and 'not equal to white clayed or muscovado'—though this lesser product was then paradoxically called 'brown muscovado'.

In 1837, the 'centrifugal' was invented in France. There called an *essoreuse* and designed originally to dry textiles, this device inspired an English patent directed at sugar in 1843, which had an immediate effect. The centrifugal produced a drier sugar, allowing it to be transported in bags for the first time rather than in the more expensive barrels. In simple terms it was a spin-dryer—using the same principle as today's washing machines—and it led to yet another leap in the quality of the sugar produced in the sugar colonies. Bags also packed together tighter, allowing steamships to carry more in their holds.

In the same year of the English patent on a sugar centrifugal, Norbert Rillieux in Louisiana introduced, if he did not originate, multiple-effect evaporation, which was more efficient than other vacuum systems. There are other claimants to the invention, including Degrand and Derosne, but Rillieux had multiple-effect evaporation systems operating in Louisiana in 1848, three years before the rival patents, and he clearly deserves a major share in the credit. Delicately termed a 'man of colour', he had trained in France, and his technique revolutionised sugar production.

While the planting, cultivation and harvesting of cane remained manual, after about 1820, even as world sugar prices plunged, steam engines began to appear, just here and there, as power sources for the mills. In 1845 the first railway in Cuba, or Latin America for that matter, was opened. From Havana to Güines, a distance of 70 kilometres, the railway could be used to transport cane much faster than horses, donkeys or cattle could haul it; this opened up the possibility of central mills servicing a number of owner-growers, a model previously discarded in a number of growing areas. The problem with a centralised mill had been that too much sugar might arrive in a short period, over-taxing the workers—but now mills could be made big enough to absorb the peaks of cane supply. More importantly, they could be powered by machines and required fewer workers.

In 1847 Earl Grey recommended that the British colonies use more central factories. The problem was commitment: most plantations were already set up with individual mills and, with capital hard to raise, nothing came of it. In 1871, when Antigua suffered hurricane damage, there was a chance for reform, but although a committee recommended a central factory system, there was no follow-up. Properties were selling so cheaply after the storm that people were able to make a profit still, even after rebuilding the small and inefficient mills.

By 1850 the vacuum pan was in wide use in other parts of the world, but Sir Henry Barkly, Governor of British Guiana, pointed out that the differential duties applied to vacuum-pan sugar by the home government had the effect of hindering the introduction of this method in the British colonies. On Java, where there were no restrictions on refining methods, 54 of the 95 contract factories used the vacuum pan in 1856, and after about 1865 it became the normal equipment in all new mills.

DEPRESSION AND BEET SUGAR

The war between beet and cane was far from over. In 1836 the best beet sugar yield was about 5.5 per cent by weight; by 1936 it was 16.7 per cent. Part of this increase came from improved plants, but more came from better methods of extraction. In 1866 Jules Robert developed a diffusion process for extracting sugar from beet, in which thin slices are treated to systematic extraction by circulating dilute juice at 90°C, leaving most of the albuminoids behind in the unruptured cells while extracting the sugar. Albuminoids carry large amounts of protein into the syrup, which must be removed. With Robert's new method, beet sugar could now be produced more cheaply. By 1880 beet sugar began to threaten cane sugar in price and volume; in 1884 the price of sugar plunged, and in 1885 the world produced more beet sugar than cane sugar. And after about the 1850s, sugar beet was growing in places like Utah, where it gained a benefit from the high cost of freighting cane sugar from anywhere else.

The problems of cane sugar did not stop there. Beet sugar could be produced as white granules, and lost no weight in transit, something the wholesalers appreciated. Beachey reports that one prominent confectioner said in 1889:

> No self-respecting confectioner will be bothered with the huge dirty casks and syrupy bags in which cane sugar is imported now; he may buy some of it for special work (as for gingersnaps) but can rarely use it without special examination.

In one market though, the whiteness of beet sugar was a drawback. In the 1890s an estimated 1500 tons of dyed beet sugar with a strong chemical smell sold as 'Yellow Demerara' every week in London. The sugar crystals were doctored to look like Demerara by adding a small amount of sulfuric acid to the clarified beet juice: the cost was only a few shillings a ton, but provided a premium of £1 or £2 a ton.

The name 'Demerara Crystals' was in fact a trade name for brownish yellow sugar coming from Demerara, later part of British Guiana, but in November 1913 a High Court appeal in London upheld the decision of a Metropolitan Police Magistrate that the name applied to a *type* of sugar, and not to its origin. One of the appeals judges argued that the prepared sugar in question was Demerara in every respect other than its origin, and that if people were offered natural sugar from Demerara, in its natural state and colour, they would probably refuse it.

As well as competition of this sort, sugar makers everywhere had to deal with taxes and bounties written to win votes and gain favours. Throughout the latter part of the nineteenth century, and in the early twentieth century, there was a continuing international brawl over sugar bounties, especially those paid by France and Germany, which were mainly paid for exported sugar. In October 1900 the Sugar Bounties Conference met once more and France, Germany and Austria agreed to end direct bounties, with France ending some of the indirect bounties. This was encouraged in part by India having placed a countervailing duty on subsidised sugar to help the Indians on Mauritius. It was also aided by the revelation that Germany's sugar cartel had been making huge profits. The French voters were now incensed that the British paid less than half of what they paid in France for the same sugar—to the benefit of the beet-growers and the refiners.

Many foreign governments, but especially America, revelled in the idea of countervailing duties, which involved placing a tax equal to the subsidy on any sugar coming from the nation subsidising their sugar. In simple terms, the nation applying the countervailing duty was able to milk a nice little cash cow at the expense of a foreign government and its taxpayers. What could be sweeter?

The sugar prices were down because there was just too much sugar around. One solution, of course, was to turn the excess into alcohol; but the European rum made from the white crystals of beet sugar lacked flavour. This did not faze the German rum makers, who sent commercial agents to scour the Caribbean for strong-flavoured rums that could be diluted as much as 7:1 with German spirits. The German rum industry prospered, and by 1914 Germany had 6000 distilleries producing 66 million gallons of alcohol per year. The First World War put an end to German ships carrying flavoursome rum from the West Indies, but that mattered little, because Germany had a greater need for all that unflavoured alcohol. It would be used to fuel the German war machine.

IMPROVING THE CANE

Sugar cane is a grass and, like other grasses, it reproduces by setting seed, while fragments that are placed in the ground will also take root. Perhaps by coincidence, perhaps not, the only variety of cane that came out of Persia into the Mediterranean lands, and then into and across the Atlantic, was a sterile form.

Over time, with growers habitually using cuttings to plant new crops and start new areas, the sugar cane had become a giant sterile clone that never produced flowers or set seed. This did not matter for production, since the cane grew so well from setts. The crop had no genetic diversity, however, and any pest that became established had a clear run, especially as sugar cane tended to be planted as almost the only crop, with fields lasting three or four years before they were retired and replanted in rotation. It was an ideal situation for wandering pests, and may explain why sugar production died out in many places.

The sterile cane, a dark slender form known as 'Creole', was the only one grown until Louis Antoine de Bougainville, the French sailor and explorer and equal of James Cook, found the 'Otaheite' cane in Tahiti in 1768. This was the cane which Cook used to make beer soon afterwards. Bougainville took samples to Mauritius (then called Ile de Bourbon) in 1768, where the cane was named 'Bourbon'. Around 1780, someone named Cossigny (probably Joseph-François Charpentier de Cossigny de Palma) brought more of this cane to Mauritius and Réunion, and cane from Java reached these islands at about the same time. By 1789, the new canes had reached the French West Indies.

The 'Bourbon' cane was carried to St Vincent in the West Indies with William Bligh in 1793, after a delay caused by a certain Fletcher Christian. A single plant reached Jamaica in 1795, and more specimens arrived in 1796. 'Batavian' cane reached Martinique in 1797, and from there reached Louisiana in 1818, where it was the dominant variety to about 1900.

From the 1790s until the 1890s 'Bourbon' was the principal cane variety grown in the West Indies. It was dropped only when there was an outbreak of the red rot disease that Buddha had referred to as *manjitthika*, and it was replaced by resistant strains of lower yield. These 'native canes' were variants collected from various parts of the world, often of uncertain provenance, going by names such as 'Transparent', 'Tanna' and 'Cheribon', and identified mainly by their rind colours.

Cane ranges in diameter from 12.5 mm to 50 mm (½ inch to 2 inches); it has regular nodes along its length, and is covered in a tough rind that surrounds an inner fleshy pith where the sugar is stored, though the 'Bourbon' or 'Otaheite' cane has a softer rind, making it better for chewing. The original cane taken to the Mediterranean and then to the New World appears to have been a hybrid of *Saccharum barberi* and *S. officinarum*, as we now understand it. The *Saccharum officinarum* described by Linnaeus in 1753 was probably the 'Creole' cane.

The 'Tanna' cane was first seen by James Cook at the Pacific island of Tanna in the southern New Hebrides in 1774, in a garden he described in his journal as 'laid out by line, abounding with plantains, sugar-canes, yams, and other roots, and stocked with fruit-trees'. This cane was taken to Mauritius in 1870, and soon afterwards, to nearby Fiji and Hawaii. Around the world, there were now a number of varieties and the obvious next step would be to hybridise them, but it was accepted wisdom that sugar cane did not grow from seedlings.

Genetics textbooks often tell us that before Gregor Mendel's work was discovered lurking in obscure journals on library shelves in 1903, people knew nothing of breeding plants. In fact, Mendel's own work of the 1860s belies this, and William Farrer was crossing wheat strains in Australia in the 1890s, using Mendelian assumptions, quite independently of Mendel's work, which was still hidden in the libraries. Sugar growers were no different in their sophistication, and Cossigny had suggested, as early as 1780, that strains of sugar cane might be crossed to combine useful features of different varieties.

By 1790 a researcher called Tussac in Saint Domingue had described the structure of the 'Creole' sugar cane flower, but as that variety is sterile there were no seedlings to work on. In 1858 Iran Aeus saw and recognised cane seedlings growing in a field of ratoon cane on Barbados. This discovery was reported by the plantation owner in a letter to the *Barbados Liberal* in 1859 and the report was copied into the Australian press. A number of other accounts of sugar seedlings appeared in the 1860s and 1870s.

Cook's 'Tanna' cane had gone to Hawaii as 'Yellow Caledonia'. In 1888 a Dutch researcher on Java announced that he had produced seedlings from a variety known there as 'Yellow Hawaii', probably a close relative. Soon scientists in Indonesia and India were hard at work, with others following. Cane seed would never be a practical way to plant a crop, but to get characteristics from two strains into one plant, seeds were a boon.

The lesson of manuring to maintain and improve soil quality had been learned by 1900, though it was mainly added now as artificial fertiliser: A study showed that around 1900, Hawaii used £8 worth of fertiliser per acre and returned 9 tons of cane per acre; Barbados used £3 10s. worth per acre for a return of two to three tons, while the rest of the West Indies used about £1 10s. worth of fertiliser per acre for a yield of just 1.75 tons.

The lesson was clear, and the twentieth century saw serious-minded agriculturists applying serious science to sugar growing, and getting serious increases. And that was just as well, because the price of sugar kept falling.

CUBA HONEY

Good brown sugar 11 lbs., water 1 quart, old bee honey in the comb 2 lbs., cream of tartar 50 grains, gum arabic 1 oz., oil of peppermint 5 drops, oil of rose 2 drops. Mix and boil two or three minutes and remove from the fire, have ready strained one quart of water, in which a table-spoonful of pulverized slippery elm bark has stood sufficiently long to make it ropy and thick like honey, mix this into the kettle with egg well beat up, skim well in a few minutes, and when a little cool, add two pounds of nice strained bees' honey, and then strain the whole, and you will have not only an article which looks and tastes like honey, but which possesses all its medicinal proper-ties. It has been shipped in large quantities under the name of Cuba honey. It will keep fresh and nice for any length of time if properly covered.

Daniel Young, *Young's Demonstrative Translation of Scientific Secrets*, Toronto, 1861

11
LABOUR
PROBLEMS

As sugar prices continued to fall, the sugar planters needed cheap labour as well as better methods and machinery, and in the end, the surviving sugar planters were those who sought both. Cheap labour is still used in the developing world to harvest cane, but in the industrial countries, cane setts are planted in furrows by machine; cultivation, spraying, harvesting and every other process is done by machine.

With the end of slavery, growers still needed access to a poor peasant class who would grow their sugar for them, and as new areas of sugar growing developed, so it became necessary to bring peasants to those areas. The polite name given to this was 'indentured labour', but where this had once meant bringing out Celtic peasants from Britain to serve English masters, it now meant bringing 'coolies', as the Asian peasants were dismissively called, from some other part of the world.

To put matters in perspective, the conditions for travel at sea were always appalling in the age of sail. Of the 775 convicts who set out in 1787 from England for Botany Bay, 40 died before they reached Australia in early 1788, and others died soon after.

Even free travellers had a hard time of it, as Janet Schaw reported after seeing a Scottish family who were heading for Jamaica on her ship in 1774 (and interestingly, paying for their passage by the husband signing indentures). Miss Schaw thought it a bad bargain when she saw what they had to eat:

> It is hardly possible to believe that human nature could be so depraved, as to treat fellow creatures in such a manner for a little sordid gain. They have only for a grown person per week, one pound neck beef, or spoilt pork, two pounds oatmeal, with a small quantity of bisket, not only mouldy, but absolutely crumbled down with damp, wet and rottenness. The half is only allowed a child, so that if they had not potatoes, it is impossible they could live out the Voyage.

THE RETURN OF INDENTURED LABOUR

The rate of slave mortality for Barbados was regarded as normal on sugar plantations, but the figures for white indentured servants were almost as bad, and in the nineteenth century, indentured labourers who came from India suffered death at the same levels. The missionaries, deprived of slaves to emancipate, suddenly found a new cause to fight: that of people who were slaves in everything but name. The big stick they had was the level of mortality that the indentured sugar workers suffered.

A small number of Indians had arrived in Mauritius from Pondicherry in 1735. The next example of indentured labour comes in a report from 1800 of about 80 Chinese working on sugar plantations in Bengal and Bombay! Peter Cunningham, a surgeon who visited Australia several times on convict ships and owned a property in Australia, visited Port Macquarie, north of

Sydney, where he saw sugar cane growing. He recommended the use of convicts and later, Chinese workers:

> With good superintendence, convicts may be made to do quite as much as ever I saw accomplished by slaves, their labour being furnished free from any primary outlay of capital, while that of the slaves must be previously purchased, the interest upon the original price of the slave amounting to at least ten pounds annually . . . But perhaps as good a plan as any would be to establish a colony of Chinese on our shores, these being the principal sugar-growers in the Indian islands, and always ready to emigrate to any place where money is to be made.

By 'Indian islands' Cunningham meant Indonesia, where Chinese sugar growers had long been a major force. A wealthy Chinese community was established on Java by about 1400, and at some stage soon afterwards started growing sugar. Certainly the Dutch found sugar manufacture under Chinese control in the East Indies, when they began arriving in the area in 1596.

We can get an idea of the size of the East Indies industry when we consider the so-called Chinese Rebellion that took place in 1740. The Governor-General, Adriaan Valckenier, realising that the sugar industry of Java was all in Chinese hands, set out to round up and sell all of those Chinese without regular employment as slaves to Ceylon. A number of wealthy Chinese were arrested in short order. The other Chinese took up arms, and as many as 10 000 were killed before peace was restored. Around 1759, we find the first record of Chinese sugar workers at Penang off the coast of the the Malay Peninsula.

Some of the Chinese were independent travellers. One entrepreneur set up a stone mill and started to make sugar in Hawaii, on Lanai in 1802, but he left the next year. In 1832 William French put up the first lasting Hawaiian mill, operated by Chinese labour. In 1836 the mill sent the first 4 tons of sugar to the United States, and about 30 tons of molasses. The importation of workers here was minor, compared with what would soon happen.

John Gladstone, father to the later British Prime Minister, W. E. Gladstone, realised soon after emancipation that Indians could be successfully introduced into the West Indies. He retained the Calcutta firm of Gillanders, Arbuthnot and Co., which had been recruiting labour for Mauritius, to obtain workers. In January 1838 the first 414 recruits left on a five-week voyage to the Caribbean. Eighteen died on board, the other 396 arrived safely. Later, Lord Brougham (a Whig politician) would claim that 20 per cent had died on the voyage, with another 30 per cent dying in the next five weeks.

Five years later, another 98 of the survivors had died, 238 had returned to India, 60 had elected to stay, and two had disappeared from view; this gave a death rate of more than 27 per cent. Those figures caused concern in those more enlightened days, and the British government would not agree to any extension of what could be represented in some quarters as 'slavery in disguise' until the returnees reached India and had been examined. In 1844 the authorities were satisfied and 'emigration' was again allowed.

John Gladstone's Indians were described as 'hill coolies', but they were in fact Dhangars, a non-Aryan caste of nomadic agricultural labourers from Chota Nagpur. Unfortunately, later selections were far less discriminating, and members of warrior castes, untouchables and others were all mixed together.

In 1845 Lord Harris, Governor of Trinidad, set up a code of regulations for the management of the Indian workers, but the BFASS had this code disallowed. Instead of the immigrants being managed, a policy of *laissez aller* prevailed and the Indians were told they were free to move on if they wished. At the end of the first year many declined to re-engage, flocking to the towns as beggars and vagrants. From this point on, Exeter Hall in its various guises tended to be an active opponent of indentured labour, claiming that it was a form of slavery, while remaining blind to the abuses visited upon working folk at the heart of the British Empire.

THE WHITE SLAVES OF ENGLAND

Robert Sherard, the great-grandson of William Wordsworth, was a journalist and social campaigner. He could see the plight of the unfortunate as clearly as the BFASS, but he was concerned about those closer to home. The main criticism that Carlyle and others in the mid-nineteenth century had offered of Exeter Hall was that the English campaigners for social justice for slaves, like the union-crushing William Wilberforce, were oblivious of the pain and suffering under their own noses. All over England, people were working in hideous conditions, as Sherard explained in his crusading work *The White Slaves of England*.

His set of essays first appeared in *Pearson's Magazine*, informing his readers about trades that few of them would ever have heard of, like the stone nobblers, old men who broke stones in an alkali works, so that sulfur could be extracted. They were paid at eightpence the ton, and a king among stone nobblers could earn thirteen shillings a week; few earned more than eight shillings. As one stone nobbler told him, 'This is the last stage before the workhouse.'

Like Queen Elizabeth in the late sixteenth century, many of the workers in chemical factories of the late nineteenth century lost their teeth, though rather less pleasantly, as they were eaten away by the fumes and dust that they breathed and swallowed.

Sherard found that a 'salt-cake man' could be recognised any-where. His teeth, if not entirely destroyed, were but black stumps, and the effect made itself seen in less than twelve months. But it was not only the old and the infirm who were treated like this, well after most of the world's slaves had been freed. Consider another girl he saw in a factory at Cradley:

> She was fourteen by the Factory Act, by paternity she was ten. I never saw such little arms, and her hands were made to cradle dolls, yet she was making links for chain-harrows . . . Next to her was a female wisp, forging dog-chains, for which she received three farthings a piece. It was the chain which sells currently for eighteenpence. She worked ten hours a day and could 'manage six chains in the day'.

The grinding poverty we meet in the novels of Charles Dickens and the works of Gustave Doré was still a grim reality even at the close of the nineteenth century. Many had no choice: it was either work in those conditions, or perish in the workhouse, which held power over the English (and even poor foreigners) who fell into its clutches right into the twentieth century.

Grace Elizabeth ('Betsy') Jennings Carmichael was an Aus-tralian poet of considerable promise, who wrote as Jennings Carmichael. Her husband, Francis Mullis, deserted her and her three sons while they were travelling in Britain and she was forced into the Leyton workhouse in 1904, where she died. Her three sons survived in another workhouse, and outraged Australians raised a public subscription to bring them back to Australia, where they unsurprisingly changed their name to Carmichael. Two generations earlier, though, the workhouses were far worse, so perhaps things were improving slowly.

Thomas Austin was a resident of the Hendon workhouse when he fell into the laundry copper and was scalded to death in 1839, soon after Britain freed all its slaves in a welter of self-congratulation at the country's humanity. The workhouse authorities quietly buried the body, but the coroner, Dr Wakley, heard of the matter. He asked for the body to be exhumed, and declared the man had died of scalding, adding a rider to the effect that the master of the workhouse had been guilty of contributory negligence in not providing a protective railing round the copper. This was too much for the workhouse master, and he observed forcefully that the jury might have found a verdict, but had not identified the body, provoking Wakley to gain instant fame when he asked, 'If this is not the body of the man who was killed in your vat, pray, Sir, how many paupers have you boiled?'

THE MASS MIGRATIONS

During the nineteenth century, many more people travelled the seas as 'free labourers' than did so in any comparable period when the slave traders operated, or even when the Spirits stole folk away from the ports of Britain. The profits may have been less, but many ships plied a trade that was, by all accounts, out of control. The need for control can be seen in the figures for Chinese coolies travelling to Cuba—between 1847 and 1880 some 140 000 were shipped around Cape Horn. About 12 per cent died on the trip, less than 25 per cent survived Cuba, and fewer than 1 per cent ever returned to China.

Jamaica was little better: of the 4551 Indians who arrived there in 1845–47, and 507 destitute Chinese who had come from Panama at the same time, just 1491 were still working in agricultural employment by 1854, with 1762 repatriated and a further 1805 dead or disappeared. (In fairness, though, it must be mentioned that a major cholera epidemic in 1850 killed 50 000 people across the island.)

Hawaii was another but rather less damaging user of indentured labour. The brig *Thetis* brought 253 Chinese to Hawaii in 1852, and by 1898 some 37 000 had landed. The 1910 census showed 21 674 Chinese still there. A group of 148 Japanese arrived on the *Scioto* in 1868, five more following in 1882 and another 1959 in 1885. This was the start of a flood, with 176 432 Japanese arriving between 1882 and 1907, when the flow was restricted by agreement. After that, emigration exceeded immigration, but there were still 45 000 Japanese in the Hawaiian islands in 1936.

There was clearly money to be made from transporting indentured labour, and more and more people started doing it. Attempts were made to start the import of Indians to Natal in 1858 and 1859; the bill approving the practice was passed in 1860. Soon afterwards, sugar workers began to flood in from India. One of the Indians who went to Natal, but as a lawyer, not a sugar worker, was Mohandas Karamchand Gandhi. In 1893, the 24-year-old Gandhi left a lucrative law practice in Bombay to work for the rights of the Indian sugar workers in South Africa, where they were made to feel that they were remarkably second rate, and needed a firm and knowledgeable representative. Clearly, Gandhi already had a strong conscience, but the man we know as Mahatma Gandhi strengthened himself for his struggle to free India while tending to the needs of the Indian sugar workers in Natal. Once again, sugar policies had an unexpected result.

The main advantage of foreign labourers anywhere seems to have been the language barrier that tied them to a workplace, but the excuse for bringing them in from other places was usually that 'the natives won't do the work'—ironically, at the same time that Fijians were being recruited to work on other islands, Indians were being recruited for Fiji. The first Indian indentured labourers arrived in Fiji in 1879, and by 1916 a total of 68 515 had arrived. A number of these Indians were repatriated, found no place for themselves in India, and so re-emigrated to Fiji, where their descendants remain today.

Ralph Shlomowitz, an Australian academic, points out that the Natal and Fiji experiences allow us to assess the death rates reported from the various sugar plantation areas in a more balanced way:

> More generally, the importance of epidemiological factors is also shown in a consideration of the variation in the average death rate of Indian indentured workers at home and abroad in the late nineteenth and early twentieth centuries. Highest death rates occurred in the epidemiologically hostile tea estates of Assam, with its endemic malaria and cholera, and the sugar cane plantations of Malaya, with its endemic malaria; lowest death rates occurred in the relatively epidemiologically benign sugar cane plantations of Natal and Fiji, generally free of malaria and cholera. Death rates on Caribbean sugar cane plantations were higher than in Natal and Fiji as malaria was endemic in the Caribbean; the death rates in the Caribbean were lower than those in Assam and Malaya because the malaria strains in the Caribbean were much less lethal than those in Assam and Malaya.

AUSTRALIA AND THE KANAKAS

For the most part, the Exeter Hall faction did little about the recruitment of indentured labour in Asia (as opposed to its use in various places), perhaps because they felt it was better supervised in India and China. However, the London Missionary Society and various Presbyterian missionaries on South Sea islands were outspoken in their complaints about the 'blackbirders', the ships' captains who recruited labour from the islands to work in the sugar plantations of Queensland and New Caledonia (people referred to at first as 'Polynesians' and then as 'Kanakas'). It was, they asserted, no better than a slave trade. This claim is maintained today by the descendants of the Kanakas who still live in Australia and by many other Australians, but the evidence is less than clear.

Anthony Trollope had visited Demerara and Trinidad before travelling to Australia, where he saw some of the first Kanakas working on the plantations:

> Then as now there was a fear in England that these foreigners in a new country would become slaves under new bonds, and that a state of things would be produced,—less horrible indeed than the slavery of the negroes who were brought into the West Indies by the Spaniards,— but equally unjust and equally opposed to the rights and interests of the men concerned . . .
>
> Let us have no slavery in God's name. Be careful. Guard the approaches. Defend the defenceless. Protect the poor ignorant dusky foreigner from the possible rapacity of the sugar planter. But . . . be not led away by a rampant enthusiasm to do evil to all parties. Remember the bear who knocked out his friend's brains with a brickbat when he strove to save him from the fly. An ill-conducted enthusiasm may not only debar Queensland from the labour she requires, but debar also these poor savages from their best and nearest civilisation.

Trollope outlined the diet of the Kanakas; under the standard contract the daily ration was 1 lb of beef or mutton (he missed the alternative of 2 lb of fish) and another 1 lb of bread or flour, 5 oz sugar or molasses, 2 lb of vegetables which might be substituted by 4 oz of rice or 8 oz of maize, with a weekly issue of 1½ oz of tobacco, 2 oz of salt and 4 oz of soap. Commenting on this, he observes that their 'dietary is one which an English rural labourer may well envy'.

Trollope quoted figures to show that the total cost of hiring a Kanaka over a three-year contract was £75, at a time when a white labourer on weekly wages of 11s. would cost about £86 over three years. Against that, he admits that in Queensland, 15s. was the usual minimum, with sugar establishments paying white workers between 15s. and 20s. a week. He also quoted a figure of 25s. a week, including rations, for white labourers, and added: 'I was told by more than one sugar-grower that two islanders were worth three white men among the canes.'

In short, it would appear that Trollope's evidence shows that the Kanakas were underpaid, but this related mainly to the first three years, when some degree of training was needed. Kanakas seeking a second contract were generally reported to be paid rather better rates, though actual figures are hard to find. Trollope, having discussed the nature of the contract signed by the Kanakas, and the extent to which they understood it, observed in relation to some of the criticisms:

> There is not a word said here that might not be said with equal force as to the emigration of Irishmen under government surveillance from the British Isles to the British colonies,—except in this, that in regard to the poor Irishman there is seldom any contract insuring him work and food and wages immediately on his arrival. Were there any such contract he would not understand it a bit better than the islander,— who does in fact know very well what the contract ensures him.

The main claim of the missionaries was that the labourers were kidnapped by the blackbirders. Those 'engaging in the Queensland labour trade' (the same parties by their own preferred description) answered that it was a blatant lie, that the missionaries were objecting because they knew that a 'boy' would be much less amenable to their demands and strictures once he had seen something of the world. William Wawn, one of the recruiters (if we may use that as a neutral term), explained it like this:

> The returned islander, however, is a very different personage for the
> missionary to operate on. He has seen the world. He does not believe in
> offerings to the church in the shape of pigs, fowls, yams and bread-
> fruit. He knows how clergymen are regarded by the white workmen
> with whom he has come in contact . . . the missionary finds him a
> terrible stumbling-block in his path.

Certainly some of the indentured labourers in other parts of the
world had been kidnapped. A commission which travelled to
Cuba from China found that of about 40 000 Chinese who had
been shipped there, around 80 per cent had been kidnapped or
decoyed. This was not the case in the South Sea, according to
William Wawn.

Whatever the reliability of Wawn's other comments, his
points about Pidgin English are certainly valid:

> This custom of making presents to recruits' friends has been eagerly
> seized upon by our opponents as proof that we really bought the
> recruits—that the latter were slaves, probably captured in war; which
> is simply absurd. New Hebrideans never spare their enemies in battle,
> or make prisoners of the men. Slavery is unknown to them; they are not
> yet sufficiently advanced to appreciate it . . .
>
> Owing to their limited knowledge of the English language, such
> terms as 'buy,' 'sell,' and 'steal,' have a wide and comprehensive
> meaning. 'You buy boy?' is often the first question asked of a recruiter
> when he arrives at a landing-place. This simply means 'Do you wish to
> engage boys?' 'Boys,' as elsewhere, signifies men of any age. The term
> 'steal' is also frequently misunderstood. If you take away a recruit from
> his home without 'buying' or 'paying' for him,—that is, without
> making presents to his friends to compensate them for losing him,—
> they will say you 'steal' him.

From his detailed defence of the 'labour trade', it is hard to tell
whether Wawn is a plausible rogue, or a knockabout ruffian
telling a (perhaps somewhat shaded) version of the truth. He

admitted that certain traders were guilty of infringements of the rights of some of their recruits, but that in general it was only those recruiting for the French colonies who did such things. He argued that if the 'recruiters' had indeed kidnapped recruits from the islands they would never be able to go back there again, that they would be destroying their future markets, or putting their lives at risk.

Wawn also applied a kind of logic to the situation, arguing that many of the workers signed up for a second contract, and that many of those signing up for their first contract were from villages where former labourers lived. He drew attention to the presence of a government official on each boat, charged with the task of ensuring total fairness, reminding his readers that the recruiter needed to sign a £500 bond that he would not engage in kidnapping.

In all probability, both sides in the debate were somewhat at fault. There must have been times when desperate traders, faced with financial ruin, seized some unfortunates, or where a bribe encouraged the government supervisor to look the other way. It is certainly the case that many of the missionaries sent out to the islands were totally unsuited to the positions they held, and entirely untrained. The same could be said of the recruiters, and each party was very happy to blame all of their woes on the other. Against that even-handed consideration, there is a weight of tradition, among both whites and Kanakas, that the labour trade was a form of slavery. So we find *The Worker* in 1911 saying that 'Australians are not likely to submit without a protest against being treated like the Kanakas of slavery days'.

We can see from the records that large numbers of workers died periodically, a fact popularly held to be have been due to harsh treatment. However, blaming the excessive mortality on harsh treatment does not explain why the mortality was highest soon after the arrival of the islanders in Queensland. Ralph Shlomowitz has shown that the death rates declined by the year of residence, and this suggests that people who understand the germ theory and epidemiology have no need to blame harsh treatment to explain what happened.

White planters began by saying that only coloured races could work the sugar cane, but later they wondered, with cheerful racism, if Mediterranean Europeans might also be up to the task. The first 180 Portuguese arrived in Hawaii in 1878, and 30 000 had followed by 1913. Between 1907 and 1913, Hawaii saw the arrival of 6588 Spaniards, mainly from Malaga, a depressed sugar district. Australia began admitting Italian migrants to work the cane fields and other nationalities followed, coming, however illogically, even from countries like Finland. Where Australia had been a stolidly, solidly British place, we might wish to trace the eventual breakdown of the unofficial White Australia Policy to the entry of the sugar workers. It was certainly a first step in making Australia less exclusively British.

All over the world, history's greatest human mass migrations were taking place. Between 1887 and 1924, for example, Argentina had a net gain of 800 000 Italians and 1 million Spaniards, many working in the sugar industry. Many more had returned home richer than when they left.

ANZAC BISCUITS

For 48 biscuits, you will need 125 g butter, 1 tbsp golden syrup, 2 tbsp boiling water, 1½ tsp bicarbonate of soda, 1 cup rolled oats, ¾ cup desiccated coconut, 1 cup plain flour, 1 cup sugar. Melt the butter and golden syrup over a gentle heat, add mixed boiling water and bicarbonate of soda. Pour the liquid into mixed dry ingredients and blend well. Then drop teaspoonsfuls of the mixture onto a greased tray and bake in a slow oven [150°C] for 20 mins. Allow the biscuits to cool on trays for a few minutes and then remove them. The biscuits should be stored in airtight containers when cool.

The standard modern recipe of a treat for First World War soldiers
from Australia and New Zealand

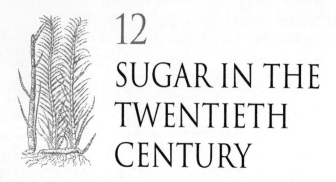

12

SUGAR IN THE TWENTIETH CENTURY

By the start of the twentieth century, sugar was a major source of energy, a major food component and a major source of agitation. Sugar proved to be an excellent source of energy for conspiracy theories, though alternative sweeteners proved to have much the same power. Where pamphleteers once wrote mainly of the evils of sugar taxes or slavery, the new pamphleteers wrote books, proclaiming the insidious nature of various sweeteners, or the promise of the new wonder fuel.

The truth was a little different.

RUNNING ON ALCOHOL

Many people are firmly convinced that the future of the internal combustion engine lies in alcohol made from sugar. This is, they argue, a clean fuel, a green fuel, a modern fuel, because while it produces carbon dioxide when it is burned, this has been taken out of the atmosphere by the sugar cane in the first place, so there is no net increase in greenhouse emissions. There is just one small problem, which becomes apparent in auditing the fuel and greenhouse costs of clearing the land (you lose the original productivity of the land), planting and cultivating the crop, making and adding fertiliser, spraying the crop, cutting and transporting the cane, crushing the cane and refining the sugar, distilling and transporting the alcohol.

Such an audit does not give quite such a rosy picture of clean, green, sugar-based alcohol fuels. Corn is slightly less efficient than sugar at providing feedstock for ethanol, but on American figures, eleven acres of corn, enough to feed seven people, would be needed to produce the fuel to take one car 10 000 miles—and the 852 gallons required would cost US$1.74 per gallon, against US$0.95 per gallon for fuel based on crude oil. To power all of the cars in the United States would use up 97 per cent of all the land area of that nation.

The first authentic internal combustion engine in America, developed by Samuel Morey around 1826, ran on ethanol and turpentine. In 1860, Nikolaus August Otto in Germany used ethanol to power an early engine—it was widely available throughout Europe as a fuel for spirit lamps. He devised a carburettor which, like Morey's, heated the alcohol to help it vaporise as the engine was being started. A January 1861 patent application on the carburettor in the Kingdom of Prussia was turned down, probably because heated alcohol carburetion was already being widely used in spirit lamps. Otto's initial financing came from Eugen Langen, who owned a sugar-refining company that probably had links to the alcohol markets of Europe.

The French—even though their grape growers were protective of the brandy market—recognised a good thing when they saw one, and French fuel-alcohol production rose from 2.7 million gallons in 1900 to 5.7 million in 1903 and 8.3 million in 1905. In a 1901 rally sponsored by the Automobile Club of Paris,

50 vehicles ranging from light quadricycles to heavy trucks made the 167-mile trek from Paris to Roubaix. After the next rally, the consumption of fuels (varying from pure alcohol to 50 per cent alcohol and 50 per cent gasoline) were measured for each vehicle. Most drivers apparently preferred the 50–50 blend.

In 1902 there was a Paris exhibition entirely devoted to alcohol-powered automobiles and farm machinery, as well as a wide variety of lamps, stoves, heaters, laundry irons, hair curlers, coffee roasters and every conceivable household appliance and agricultural engine that could be powered by alcohol. These were not experimental models but reflected a well-established industry.

Ten per cent of the engines being produced in 1906 by the Otto Gas Engine Works in Germany were designed to run on pure ethanol, while a third of the heavy locomotives produced at the Deutz Gas Engine Works ran on pure ethanol. In 1915, when oil shortages seemed likely to paralyse Germany's transportation system, thousands more engines were quickly modified to run on ethanol. In retrospect, it seems possible that Germany could have been beaten by 1917 if the production of alcohol from beet sugar had not formed an important part of the agricultural economy.

Perhaps the Kaiser gave time to thank that Prussian ancestor who had encouraged Marggraf to work with beet sugars, but perhaps those whose menfolk died in the mud and blood of Flanders would have been less appreciative of the part sugar played in prolonging a war that nobody could really win. But then again, sugar had long been killing human beings.

SUGAR AND HEALTH

Buddha said it was no sin for the sick to ask for *gur*, and from the time of the Romans through to the Middle Ages sugar was more a medical treatment than a treat. Thomas Aquinas approved of sugar being administered during Lent, but there were others with different views. By 1606 the French physician Joseph Du Chesne had already stated that sugar was essentially dangerous, that its whiteness was hiding dark perils.

On the other hand, the British physician Tobias Venner wrote in 1620 that 'sugar by how much the whiter it is, by so much the purer and wholesomer it is . . .'. Venner also argued that tobacco was health-giving, so he may not be a reliable source. In 1647 another doubtful commentator, the French-born British author Theophilus Garencières, accused sugar of being the cause of *Tabes Anglica*, later known as consumption and which we now call tuberculosis. The same author was also rather keen on 'tincture of coral' as a general cure, and used the prophecies of Nostradamus to show that King Charles II would have a son. Sadly for both prophets' reputations, Charles died childless.

According to Garencières, sugar was 'not only injurious to the lungs in its temperament and composition, but also in its entire property', a view that found support from Dr Thomas Willis, commonly credited with the discovery of glycosuria in 1674, who not only quoted Garencières favourably, but expressed the opinion that in addition to tuberculosis, sugar was responsible for scurvy.

In France too, sugar was under attack. Philippe Hecquet, at one time physician to Louis XIV, was associated with the Jansenists, an influential 'puritan' sect within Catholicism. He was adamant that sugar was an essentially treacherous substance, a poison that used pleasure as a lure. In fact, the theme of sugar as lure continues right up to the present day. In a similar way, the different materials used in sugar making have also come under attack. When Joseph Banks discussed the poor state of the teeth of the betel-nut chewers of Savu in 1770, he blamed the lime used in refining sugar:

> This loss of the teeth is . . . in my opinion is much easier accounted for by the well known corrosive quality of the lime, which is a necessary ingredient in every mouthfull and that too in no very insignificant quantity . . . Possibly the ill effects which sugar is beleivd by us Europeans to have upon the teeth may proceed from the same cause as it is well known that refin'd or loaf sugar contains in it a large quantity of lime.

In the late nineteenth century the wheel turned again, and many physicians in France and Germany were loud in their praise of the health-giving properties of sugar. Opinion changed again in the twentieth century, and is negative to this day, with websites warning solemnly of danger, like this example:

> In the process of making sugar from both cane and beets, they are heated and calcium hydroxide (lime), which is a toxin to the body, is added. This is done to remove those ingredients that interfere with the complete processing of sugar. Carbon dioxide, which is another toxin, is then used to remove the lime (and according to my studies, not all is removed).

This bizarre claim of remnant carbon dioxide in the sugar is at odds with the mainstream terror campaign that sugar is 'pure, white and deadly', to use the tag line of John Yudkin. William Dufty warns in a best-selling book that:

> So effective is the purification process which sugar cane and beets undergo in the refineries that sugar ends up as chemically pure as the morphine or the heroin a chemist has on the laboratory shelves. What nutritional virtue this abstract chemical purity represents, the sugar pushers never tell us.

To Dufty, purity offers all sorts of magical powers, and apparently even the hint of purity in the future can work wonders, for he argues that sugar caused madness at a time when it was far from pure:

In the Dark Ages, troubled souls were rarely locked up for going off their rocker. Such confinement began in the Age of Enlightenment, after sugar made the transition from apothecary's prescription to candymaker's confection. 'The great confinement of the insane', as one historian calls it, began in the late 17th century, after sugar consumption in Britain had zoomed in 200 years from a pinch or two in a barrel of beer, here and there, to more than two million pounds per year.

This sort of argument is known among logicians as *post hoc, ergo propter hoc*, a Latin tag that means 'after that, so caused by that'. On this principle, if I die after drawing breath, I died of a surfeit of air; if I die on exhaling, I died because I failed to hold my breath. In short, it turns a chance correlation into a major issue. It is as invalid as it would be for me to say I lost weight, my haemorrhoids disappeared, and I felt better after avoiding sweets, while neglecting to mention that I also gave up cocaine and started exercising at the same time.

Correlations are the stock in trade of people seeking to make a case: while it may be true, for example, that some Presbyterian ministers in Massachusetts drank a lot of rum, the linear relationship between their average salaries and the price of rum sold in Havana over several centuries is unlikely to be any more than the result of comparing two values both influenced by the cost of living. And if correlations are evidence, when there were wireless licences in Britain there was a strong correlation between the number of admissions to British mental institutions and the number of licences issued for radio receivers.

Eventually people started to come out with all sorts of fearful tales based on the purity of sugar which—by its very purity—shocked the digestive system as no food had ever done in the past! In truth, as soon as sugar is mixed with other foods, or with saliva and digestive juices, it ceases to be pure—end of story. In the 1970s the Australian sugar industry did not try to fight fads with facts: they simply advertised that sugar was natural, and restored market share. This soothed the minds of consumers, who neglected to consider that hemlock, lightning, plummeting asteroids, strychnine, cowpats, poison ivy, arsenic, great white sharks, stinging nettles, tarantulas, cobras and skunks are all natural as well. The argument worked well, but the science was as weak as the science in the attacks on sugar—but that was nothing to the attacks on the other sweeteners.

The alternatives

The alternative sweeteners, just like sugar, seem to attract their share of less than entirely rational opposition. They are accused of causing diseases, breaking down to produce poisons and, worst of all, they are accused of having been discovered by accident. To people with a low level of scientific knowledge, like the food faddists, accidental discoveries just have to be the work of the Devil, or worse.

The problem with food faddists is that if they go on long enough, then like the infinite number of monkeys that I have currently working on the sugar-free version of *The Winter's Tale*, they will hit on the truth at some point. But before they get there, they will have raised so much noise and clatter that the small medical truth, when it is found, may be missed in the general hullabaloo. In fact many scientific discoveries may be attributed to a chance event, and attacking the end product because it was found by accident is far from the rational end of the criticism scale. Accidents are far more common in the chemical laboratory than outsiders realise!

Accident 1

Constantine Fahlberg, who patented saccharin in 1879, claimed that he accidentally spilled some chemical on his hand while working on possible food preservatives in Ira Remsen's laboratory at Johns Hopkins University. Fahlberg noticed a sweet taste while eating dinner that night, and then sought it out in the laboratory the next day. Fahlberg and Remsen published a joint paper on the new substance, but then Fahlberg took out the patent on his own, and became rich. Remsen went on to become president of Johns Hopkins, and is reported to have said later, 'Fahlberg is a scoundrel. It nauseates me to hear my name mentioned in the same breath with him.'

Saccharin was a wonder in its time, and before the First World War, when it was taxed in terms of its sweetening power (rated then at 500 times the sweetness of sugar), quite a lot of saccharin was smuggled into Britain. Sir William Tilden describes the checking of a number of 'white powders':

> . . . the inducement to smuggle this article into the country is very great, and numerous ingenious methods have been devised for this purpose have been detected by the Customs Department. The presence of saccharin, therefore, has to be searched for in all preparations in which there is any probability of its occurrence. Saccharin was discovered in 83 samples specially examined . . . in the year 1911–12.

In fact, saccharin was used much more commonly in Europe after the Americans joined in the First World War and, with sugar rationed, the little pink packets became common. Since 1960 there has been some evidence of bladder cancer in rats dosed with saccharin, but no human cancers have been associated with the product. Saccharin has been banned in Canada for many years, but not in the United States, producing a new generation of saccharin smugglers motivated more by matters relating to diabetes than to evading excise and customs.

Accident 2

Michael Sveda, a graduate student at the University of Illinois in 1937, was investigating the structures of antipyretics, drugs that reduce fevers. While studying sulfamates and their derivatives, he produced cyclamate, but at this point chemical legend and the story Sveda told a few years later diverge. The legend has it that he picked a cigarette up off the bench, where it had rolled in some spillage. He was indeed smoking in the laboratory, as happened in those more innocent days, but he merely lit the cigarette without washing his hands, and so transferred a small sample of the compound to his mouth.

His son, also a scientist, recalls that when cyclamates were released commercially Sveda was asked to endorse a brand of cigarette, but he declined the opportunity. An intensely honest man by all accounts, he was later wounded by the campaign that was directed at cyclamate/saccharin sweeteners, based on a bizarre experiment that induced cancer in rats when they were given a dosage equal to a human drinking a bathtub of cyclamate-sweetened drink, every day for a year.

Another public demonstration showed deformed chickens which had been injected with cyclamates, when injections of salt, water or even air would probably have had the same effect. These were stunts aimed at stampeding the American public, and it is not too hard to see which part of the marketplace might gain from such a scare campaign. Not that I am making any such allegation, but it bears thinking about.

The science journal *Nature* commented after America banned cyclamates that 'it would be all too easy for public apprehension to be raised to the pitch where a fever of vegetarian faddism drives everything but mothers' milk from the market'. Others called it bad science, and strangely done, but it was enough to bring in a ban in the United States, although Britain, Australia, New Zealand and Canada all allow the product in some forms, as do many other countries. The American ban persists, and to this day nobody can quite see why it was imposed, or why it is maintained.

Accident 3

James Schlatter's accident came in 1965 while he was studying compounds that might be useful against ulcers. The target was a tetrapeptide, a chain of four of the amino acids that are found in all proteins, which was being made from two dipeptide intermediates. One of these, aspartyl-phenylalanine ester, is what we now call aspartame. Schlatter accidentally spilt a few drops of aspartame on his hand; when he later licked his finger to pick up a piece of paper, he tasted an intensely sweet taste. Schlatter and his colleague Harman Lowrie, knowing that the compound contained nothing that did not occur in other proteins, tried a small amount of the chemical in black coffee, noted the absence of a bitter after-taste, and wrote up their discovery.

Aspartame is currently under attack because of the breakdown products it supposedly yields. In particular, it is a popular target for those claiming Gulf War syndrome problems, since aspartame breaks down at high temperatures. There is nothing in the literature of science to back this at the moment, but claims such as this, along with assertions that aspartame is 'RNA-derived', are generally thought to be enough to damn it forever.

Accident 4

The most amazing tale is also the most difficult to confirm, and it is still unconfirmed. In this story, Shashikant Phadnis, an Indian student in London, was working on halogenated sugars, which are sugar molecules with chlorine atoms substituted. He was told to *test* a compound, misheard this as an instruction to *taste* it, did so, and discovered sucralose.

All my attempts to track down any of those involved at the time have proved fruitless. There is no trace of any Shashikant Phadnis on the Internet, but a person of this name appears on the patent for sucralose. Tate and Lyle, who funded the research and who profit from sucralose, do not answer any queries that are directed at them. One can only conclude that the public relations department at Tate and Lyle maintains a fond belief that the questioner will go away and say nothing about this.

The problem with sweeteners is that there are always scare campaigns going on, and just as those two interest groups known as the East India Plunderers and the West Indies Floggers misused political power in eighteenth- and nineteenth-century Britain for economic advantage, so it seems some people have misused science for economic advantage. Most of the 'science' we hear about rival sweeteners seems to have come from the spin doctors, rather than from the medical doctors.

OIL PASTE BLACKING

Take oil vitriol, 2 ozs., tanners oil, 5 ozs., ivory black, 2 lbs., molasses, 5 ozs; mix the oil and vitriol together, let it stand a day, then add the ivory black, the molasses, and the white of an egg; mix well, and it is ready for use.

Daniel Young, *Young's Demonstrative Translation of Scientific Secrets*, Toronto, 1861.

EPILOGUE
THE COSTS AND
BENEFITS

Much of the history of sugar of the past four centuries seems to involve the jostlings of various European countries: the Dutch with the Spanish, Portuguese, English and French, Germany with everybody, England with America and, most famously, England with France. The history of sugar also involves the predation of those countries on the nationals of many other lands, either by deliberate enslavement or in the more polite guise of indentured servitude, barely a notch above slavery.

In these more enlightened times sugar slaves are often replaced by machines, but there are still parts of the world where the back-breaking task of hand-harvesting sugar cane goes on, and the lot of the sugar worker today is little better than that of the slave of yesteryear.

Sugar has caused the mass movement and death of millions of humans. It has resulted in the large-scale clearance of land and the destruction of soil and whole environments. On the plus side, it has provided us with many taste delights, and had a beneficial effect on the economies of many nations.

Sir Eric Williams, a Marxist scholar and Trinidad's first Prime Minister, argued first in his doctoral thesis, and later in his book *Capitalism and Slavery*, published in 1944, that the sugar and slave trade had provided the pump-priming, the finance, for the Industrial Revolution in Europe, although this is now generally regarded as an extreme viewpoint. Sugar slavery probably did no harm to the European economy in providing working capital in Britain and France, but Germany managed to industrialise without colonies, slaves or cane sugar.

Sugar was never the friend of socialism. When the British Labour government planned in 1949 to nationalise the sugar industry, the main British producer, Tate and Lyle, fought back, sending sugar out in bags labelled 'Tate not State'. Sugar had a great effect on Marxist Cuba after Fidel Castro's 1969 announcement that the country was to aim for a 10 million ton sugar harvest in 1970, to earn money to fund industrialisation. While the harvest was a record at 8.5 million tons, the sugar-led recovery failed, for the Cuban economy was in tatters because of the over-emphasis on one product. Cuba returned after that to a more normal Marxist economy based on the Russian model.

One of the most unusual effects of sugar is seen in the Great Boston Molasses Flood of 15 January 1919. According to some accounts, there had been a rush to import as much molasses as possible in order to make and sell as much rum as possible before Prohibition came into force. The flood was caused by an over-filled storage tank bursting and flooding the streets with a two and a half metre wave of molasses that killed 21 people, crumpled the steel support of an elevated train, and knocked over a fire station. The story has often been told since, usually with a wry comment about the victims meeting a sticky end. It is another example of the way sugar products can be dangerous, though the main problems have been environmental.

Sugar cane is a tropical crop and, like coral animals, does best in the tropics. In Florida, the Caribbean, Australia and Mauritius, mangrove swamps, seagrass beds and coral reefs are all under threat from increased sediment and fertiliser run-off from the fields of sugar cane, moving down the rivers and into the sea. The main nurseries supporting new generations of life in the sea are under serious threat because of the way sugar is grown on the land. Part of the problem is that consumers do not pay anything like the true cost of sugar, any more than they did in the days of slavery. The costs of environmental destruction cannot be readily converted into dollar amounts, and so they are largely ignored.

Yet there are solutions, and some of them are simple, like 'green cane trash blanket harvesting', which leaves a cover of organic material on the cane fields that minimises sediment movement, which can be significant during heavy rainfall. This can be done because mechanical harvesting does not require that the cane fields be burnt as they were previously to prevent the cane-cutters contracting leptospirosis, the germs of which are left on the cane by rats. If the cane stalks are left on the soil after harvest as a trash blanket, the headlands, the strips around the outside of the cane fields that are used for machinery access, will be the only remaining major sediment source.

All farming results in a steady run-off of pesticides and fertiliser, but the location of the cane fields makes them a particular problem. In some places the environmental damage is even worse. Liquid waste from sugar mills and refineries, material called *vinasse*, is discharged directly into the sea on Guadeloupe, for example, causing a serious breakdown of the marine environment. If we need sweetness in our food, perhaps we should seek other ways of finding it, because right now our joint human sweet tooth looks set to cause a nasty abscess in the environment.

ÖSTERMALMSGLÖGG

500 mL vodka, 330 mL strong beer, 750 mL Madeira, 3 dried figs, 150 grams raisins, 20 peeled almonds, 3 bitter orange peel, 1 cinnamon stick, ginger, 300 mL sugar, 6 cloves, 10 cardamons. Chop the figs into four pieces. Mix all spices except the sugar, add the beer and boil it. Add the sugar and boil until the sugar has melted. Let it cool, then add the vodka, and reheat gently—no more than 55°C (130°F) or the vodka will evaporate. Let it stand for an hour, add the Madeira, and let it stand for another hour. Pour the glögg through a mesh (but keep almonds and raisins) into bottles. Warm to no more than 55°C before serving.

Swedish recipe (modern)

GLOSSARY

bagasse the fibrous material left over after sugar cane has been through the rollers, and all the juice has been squeezed out.

claying a method of trickling water through raw sugar to whiten it.

creole/Creole depending on the context, a *lingua franca* that develops where cultures meet; a member of the plantocracy of some Caribbean islands; a person of mixed racial origins; the sterile sugar cane used in the Mediterranean which was taken to the Americas.

defecant material added to the boiling sugar juice to precipitate impurities.

eddoes a form of food found originally on the Gold Coast of Africa, and taken to the New World. It is used for the root of various plants, including *Colocasia esculenta* and *Colocasia macrorhiza* (taro).

guinea grass a grass used as feed for the animals that operated mills and hauled cane.

gur a sticky sweet ball of dark sugar, obtained by boiling cane juice.

isinglass   a form of gelatin, originally made from the swim bladders of sturgeons, and sometimes used to clear sugar juice.

Jamaica train   a system of pans used to heat sugar juice over a single fire, with the juice being moved progressively down to the final pan. It was known to the French as 'the English train' and to the Cubans as 'the French train'.

lime   calcium oxide, prepared by lime burners, who roasted shells, coral or marble.

molasses   the sticky brown liquid that is left after sugar crystals have been taken from a batch of juice.

ratoon   a later crop of sugar cane, growing from previously planted roots left in the ground after an earlier crop has been cut. Usually the cane was replaced after two or three ratoon crops.

rind   the outer layer of the sugar cane, surrounding the soft pith that actually contains the sugar juice. The forms cultivated in the Pacific, where they were used mainly for chewing, generally had been selected for a soft rind.

sett   a cut piece of sugar cane, usually with two or three joints, used to start new cane growth.

three-roller mill   the form of mill which was most efficient in crushing the cane and extracting the juice.

vinasse   the fluid left after rum has been distilled from the fermented sugar juice, and which can be a major pollutant of rivers if it is dumped.

REFERENCES

Some of the authors listed here have been directly cited, others have merely been consulted. In some cases the facts I have given may contradict one or more of the sources. This will be because the references themselves have been at odds, and I have elected to trust one source over another.

Agnew, John R. (ed.), *Australian Sugarcane Pests*. Indooroopilly: Bureau of Sugar Experiment Stations, 1997.

Allen, Richard Blair, *Slaves, Freedmen, and Indentured Labourers in Colonial Mauritius*. Cambridge: Cambridge University Press, 1999.

Anonymous pamphlet, *Case of the Refiners of Sugar in England*, c. 1695. Original in the Goldsmiths'-Kress library of economic literature, microfilm copy at the State Library of NSW.

Anonymous pamphlet, *Reasons Against Laying an Additional Duty on Muscovada Sugar*, c. 1695. Original in the Goldsmiths'-Kress library of economic literature, microfilm copy at the State Library of NSW.

Anonymous pamphlet, *Reasons Humbly Offer'd, Why a Duty should not be Laid on Sugars*, c. 1744. Original in the Goldsmiths'-Kress library of economic literature, microfilm copy at the State Library of NSW.

Anonymous pamphlet, *The Irregular and Disorderly State of the Plantation Trade*, c. 1695. Original in the Goldsmiths'-Kress library of economic literature, microfilm copy at the State Library of NSW.

Aykroyd, W. R., *Sweet Malefactor*. London: Heinemann, 1967.

Banks, Sir Joseph, *The Endeavour Journal*, edited by J. C. Beaglehole. Sydney: Angus and Robertson, 1962, now more readily available online as text.

Barnes, A. C., *The Sugar Cane*. Aylesbury: Leonard Hill Books, 1974.

Barton, G. B., *History of New South Wales from the Records, Volume I*. Sydney: Charles Potter, Government Printer, 1889.

Beachey, R. W., *The British West Indies Sugar Industry in the Late 19th Century*. Oxford: Blackwell, 1957.

Beckles, Hilary McD., *White Servitude and Black Slavery in Barbados, 1627–1715*. Knoxville: The University of Tennessee Press, 1989.

Beeton, Mrs Isabella, *The Book of Household Management*. London: S. O. Beeton, 1861.

Benvenisti, Meron, *The Crusaders of the Holy Land*. Jerusalem: Israel Universities Press, 1970.

Boswell, James, *Life of Johnson*. London: Also available as Project Gutenberg file ljnsn10.txt.

Burgh, N. P., *A Treatise on Sugar Machinery.* Bucklersbury: E. and F. N. Spon, 1863.

Carlyle, Thomas, *Occasional Discourse on the Negro Question*, originally published in *Fraser's Magazine for Town and Country*, December 1849. Print copies are hard to locate, but it can be found on the Web.

Clark, Ralph: see Fidlon, Paul G., and R. J. Ryan, *The Journal and Letters of Lt. Ralph Clark.*

Codrington, Edward, *Memoir of the Life of Admiral Sir Edward Codrington*. London: 1873.

Cook, James, *Captain Cook's Voyages of Discovery*. Everyman, 1906.

Crowe, Mitford, quoted in Noël Deerr, *The History of Sugar*, 413.

Cunningham, Peter, *Two Years in New South Wales* (first published 1827). Sydney: Angus & Robertson, 1966.

Darwin, Erasmus, *The Botanic Garden*. London: J. Johnson, 1799.

Deerr, Noël (compiled by John Howard Payne), *Noël Deerr: Classic Papers of a Sugar Cane Technologist*. Amsterdam, New York: Elsevier Science Publishing Co., 1983.

Deerr, Noël, *The History of Sugar*. London: Chapman and Hall, 1949–50.

Denoon, Donald, et al. (ed.), *The Cambridge History of the Pacific Islanders*. Cambridge: Cambridge University Press, 1997.

Duffy, Michael, *Soldiers, Sugar, and Seapower: The British Expeditions to the West Indies and the War against Revolutionary France*. Oxford: Clarendon Press, 1987.

Dufty, William, *Sugar Blues*. Tunbridge Wells: Abacus Press, 1980. (Author is also cited as Duffy, and appears to have used this name as well.)

Duhamel du Monceau, Henri Louis, *L'art de reffiner le sucre*. Paris: 1764.

Dutton, Geoffrey, *The Hero as Murderer: The Life of Edward John Eyre*. Ringwood: Penguin Books, 1977. (First published by Collins, 1967.)

Edwards, Bryan, *History of the British Colonies in the West Indies*. London: John Stockdale, 1793.

Fidlon, Paul G., and R. J. Ryan, *The Journal of Arthur Bowes Smyth*. Sydney: Australian Documents Library, 1979.

Fidlon, Paul G., and R. J. Ryan, *The Journal and Letters of Lt. Ralph Clark*. Sydney: Australian Documents Library, 1981.

Fitzgerald, Ross and Hearn, Mark, *Bligh, Macarthur and the Rum Rebellion,* Kangaroo Press, Sydney, 1988.

Flannery, Tim, *The Future Eaters*. Port Melbourne: Reed Books, 1994.

Flannery, Tim, *The Eternal Frontier*. Melbourne: Text Publishing Company, 2001.

Galloway, J. H., *The Sugar Cane Industry: An Historical Geography from its Origins to 1914*. Cambridge: Cambridge University Press, 1989.

Gilfillan, George, *Specimens with Memoirs of the Less-Known British Poets*. Edinburgh, three volumes, 1860.

Gilmore, John, *The Poetics of Empire: A Study of James Grainger's 'The Sugar-Cane'*. London: The Athlone Press, 2000. (This volume includes the complete text of James Grainger's 'Sugar-Cane: A poem in four books', first published in 1764.)

Grainger, James: see Gilmore, John.

Hentzner, Paul, *Paul Hentzner's Travels in England, during the Reign of Queen Elizabeth*. Published in London, 1797; available as an electronic text from Project Gutenberg.

Herodotus, *The Histories, Book 4*. Harmondsworth: Penguin Classics, 1954.

Higham, Charles, *The Bronze Age of Southeast Asia*. Cambridge, Melbourne: Cambridge University Press, 1996.

Hobhouse, Henry, *Seeds of Change: Five plants that transformed mankind*. London: Sidgwick and Jackson, 1985.

Huetz de Lemps, Alain, *Histoire du Rhum*. Paris: Editions Desjonquères, 1997.

James, C. L. R., *The Black Jacobins,* 2nd edition. New York: Vintage Books, 1963.

Johnston, W. Ross, *A Documentary History of Queensland*. Brisbane: University of Queensland Press, 1988.

King, Norman J., Mungomery, R. W. and Hughes, C. G., *Manual of Cane-growing*. Sydney: Angus and Robertson, 1953.

Koike, Hideo, *Sugar-Cane Diseases: A Guide for Field Identification*. Rome: FAO, 1988.

Kurlansky, Mark, *Cod*. London: Vintage Books, 1999.

Landman, Captain, *Adventures and Recollections of Captain Landman*, 1852, as quoted in Barton, G. B., op. cit., p. 498. Sydney: Charles Potter, Government Printer, 1889.

Lewis, D. B. Wyndham, and Charles Lee, *The Stuffed Owl: An Anthology of Bad Verse*. London: J. M. Dent and Sons, 1948.

Ligon, Richard, *A True & Exact History of the Island of Barbados*. London: Humphrey Moseley, 1657. (Ann Arbor microfilm.)

Littleton, Edward, *The Groans of the Plantations*. London: M. Clark, 1689. (Ann Arbor microfilm.)

Mandeville, Sir John (pseudonym), *The Travels of Sir John Mandeville*, accessed as Project Gutenberg file tosjm10.txt.

Markham, Gervase, *The English House-wife*. London: Printed by Anne Griffin for Iohn Harrison, as the Golden Vnicorne in Pater-noster-row, 1637, in microform (first edition 1615).

Matra, James Maria, "A Proposal for Establishing a Settlement in New South Wales" in Barton, op. cit., 423–9.

McDonald, Roderick A., *The Economy and Material Culture of Slaves*. Baton Rouge: Louisiana State University Press, 1993.

Meyer, Jean, *Histoire du Sucre*. Paris: Editions Desjonquères, 1989.

Mill, John Stuart, *The Negro Question*, originally published in *Fraser's Magazine for Town and Country*, January 1850. Available on the Web, usually with Carlyle's piece, to which it is a reply.

Montagu, Lady Mary Wortley, letter quoted in John Carey (ed.) *The Faber Book of Science*, 51. London: Faber and Faber, 1995.

Nabors, Lyn O'Brien, and Robert C. Gelardi (ed.), *Alternative Sweeteners*. New York: M. Dekker, Inc., 1986.

Parkinson, C. Northcote, *Trade in the Eastern Seas Between the Years 1793–1813*. London: Cass, 1966.

Phillip, Arthur, *The Voyage of Governor Phillip to Botany Bay*. London: John Stockdale, 1790.

Polo, Marco, *The Travels of Marco Polo*. New York: Liveright, 1926. London: Routledge & Kegan Paul, 1931.

Sambrook, Pamela, *Country House Brewing in England, 1500–1900*. London: Hambledon Press, 1996.

Schaw, Janet, *Journal of a Lady of Quality*. Yale University Press, 1921, with several editions. (Written in 1774.)

Sherard, Robert Harborough, *The White Slaves of England*. London: James Bowden, 1897.

Shlomowitz, Ralph, *Marx and the Queensland Labour Trade*. Working Papers in Economic History, No. 54, May 1992. Adelaide: Flinders University.

Shlomowitz, Ralph, *Marx and the Queensland Labour Trade: A Further Comment*. Working Papers in Economic History, No. 66, August 1995. Adelaide: Flinders University.

Smith, Adam, *The Wealth of Nations*. London: 2nd edition, W. Strachan and T. Cadell, 1778.

Smith, Anthony, *Blind White Fish in Persia*. London: George Allen and Unwin, 1953 (also Penguin, 1990).

Smith, Dudley, *Cane Sugar World*. New York: Palmer Publications, 1978.

Strong, L. A. G., *The Story of Sugar*. London: Weidenfeld and Nicholson, 1954.

Tilden, Sir William, *Chemical Discovery and Invention in the Twentieth Century*. London: Routledge, c. 1916.

Tobin, James, *A Plain Man's Thoughts on the Present Price of Sugar*. London, J. Debrett, c. 1792. Original in the Goldsmiths'-Kress library of economic literature, microfilm copy at the State Library of NSW.

Tomich, Dale W., *Slavery in the Circuit of Sugar: Martinique and the World Economy, 1830–1848*. Baltimore: Johns Hopkins University Press, 1990.

Trollope, Anthony, *Australia*. First published 1873 as *Australia and New Zealand*, reprinted in part as *Australia*. Brisbane: University of Queensland Press, 1967.

Unwin, Rayner, *The Defeat of John Hawkins*. London: George Allen and Unwin, 1960.

Walpole, Horace (trans.), *Itinerariae Germaniae, Galliae, Angliae et Italiae* (*Paul Hentzner's Travels*). London: Edward Jeffrey, 1797.

Ward, J. R., *British West Indian Slavery*, 1750–1834. Oxford: Clarendon Press, 1988.

Watson, Andrew M., *Agricultural Innovation in the Early Islamic World*. Cambridge, New York: Cambridge University Press, 1983.

Wawn, William T., *The South Sea Islanders and the Queensland Labour Trade,* originally published in 1893, reprinted in 1973, edited and annotated by Peter Corris, Canberra: Australian National Univesity Press.

Whelan, Elizabeth, 'The bitter truth about a sweetener scare', *Wall Street Journal*, 26 August 1999.

Williams, Eric, *Capitalism and Slavery*. London: Deutsch, 1964; University of North Carolina Press, 1994 (reported as first published 1944).

Williamson, Edwin, *The Penguin History of Latin America*. Penguin Books, 1992.

Young, Daniel, *Young's Demonstrative Translation Of Scientific Secrets*, Toronto: Roswell and Ellis, 1861.

INDEX

'Sir, my friend John Baynes used to say that the man who published a book without an index ought to be damned ten miles beyond Hell, where the Devil could not get for stinging nettles.'

Francis Douce, also attributed to Carlyle